Effluents from Alternative Demilitarization Technologies

NATO Science Series

A Series presenting the results of activities sponsored by the NATO Science Committee. The Series is published by IOS Press and Kluwer Academic Publishers, in conjunction with the NATO Scientific Affairs Division.

General Sub-Series

A. Life Sciences	IOS Press
B. Physics	Kluwer Academic Publishers
C. Mathematical and Physical Sciences	Kluwer Academic Publishers
D. Behavioural and Social Sciences	Kluwer Academic Publishers
E. Applied Sciences	Kluwer Academic Publishers
F. Computer and Systems Sciences	IOS Press

Partnership Sub-Series

1. Disarmament Technologies	Kluwer Academic Publishers
2. Environmental Security	Kluwer Academic Publishers
3. High Technology	Kluwer Academic Publishers
4. Science and Technology Policy	IOS Press
5. Computer Networking	IOS Press

The Partnership Sub-Series incorporates activities undertaken in collaboration with NATO's Partners in the Euro-Atlantic Partnership Council – countries of the CIS and Central and Eastern Europe – in Priority Areas of concern to those countries.

NATO-PCO-DATA BASE

The NATO Science Series continues the series of books published formerly in the NATO ASI Series. An electronic index to the NATO ASI Series provides full bibliographical references (with keywords and/or abstracts) to more than 50000 contributions from international scientists published in all sections of the NATO ASI Series.
Access to the NATO-PCO-DATA BASE is possible via CD-ROM "NATO-PCO-DATA BASE" with user-friendly retrieval software in English, French and German (© WTV GmbH and DATAWARE Technologies Inc. 1989).

The CD-ROM of the NATO ASI Series can be ordered from: PCO, Overijse, Belgium.

1. Disarmament Technologies – Volume 22

Effluents from Alternative Demilitarization Technologies

edited by

Francis W. Holm
Placitas,
New Mexico, U.S.A.

Kluwer Academic Publishers

Dordrecht / Boston / London

Published in cooperation with NATO Scientific Affairs Division

Proceedings of the NATO Advanced Research Workshop on
Effluents from Alternative Demilitarization Technologies
Prague, Czech Republic
October 13–15, 1997

A C.I.P. Catalogue record for this book is available from the Library of Congress.

ISBN 0-7923-5253-X

Published by Kluwer Academic Publishers,
P.O. Box 17, 3300 AA Dordrecht, The Netherlands.

Sold and distributed in North, Central and South America
by Kluwer Academic Publishers,
101 Philip Drive, Norwell, MA 02061, U.S.A.

In all other countries, sold and distributed
by Kluwer Academic Publishers,
P.O. Box 322, 3300 AH Dordrecht, The Netherlands.

Printed on acid-free paper

Printed in the Netherlands

TABLE OF CONTENTS

PREFACE

FRANCIS W. HOLM
30 Agua Sarca Road, Placitas, New Mexico

1. Overview

The North Atlantic Treaty Organization (NATO) sponsored an Advanced Research Workshop (ARW) in Prague, Czech Republic, on October 13-15, 1997, to collect and study information on effluents from alternative demilitarization technologies and to report on these findings. The effluents, or process residues, identified for assessment at the workshop are generated by systems that have been proposed as alternatives to incineration technology for destruction of munitions, chemical warfare agent, and associated materials and debris.

The alternative technologies analyzed are grouped into three categories based on process bulk operating temperature: low (0-200 C), medium (200-600 C), and high (600-3,500 C). Reaction types considered include hydrolysis, biodegradation, electrochemical oxidation, gas-phase high-temperature reduction, steam reforming, gasification, sulfur reactions, solvated electron chemistry, sodium reactions, supercritical water oxidation, wet air oxidation, and plasma torch technology. These categories represent a broad spectrum of processes, some of which have been studied only in the laboratory and some of which are in commercial use for destruction of hazardous and toxic wastes. Some technologies have been developed and used for specific commercial applications; however, in all cases, research, development, test, and evaluation (RDT&E) is necessary to assure that each technology application is effective for destroying chemical warfare materiel.

Table 1 contains a list of more than 40 technologies from a recent report for the U.S. Army [1]. Many of the technologies in Table 1 are based on similar principles. The technologies presented in bold face type have demonstrated some potential for chemical demilitarization applications and are being evaluated by the US National Research Council and the US Army for possible implementation.

A few low-temperature, low-pressure technologies have the capability to provide for destruction in a single-step process and even fewer can be expected to be applied successfully to cleaning metal parts and destroying energetics in chemical weapons. In most cases, the application requires a two-step process, combining technologies in series, to achieve the level of destruction required. The discussion at the workshop placed heavy emphasis on the ability of these low-temperature technologies to detoxify agent, metal, and energetics, producing products of reduced toxicity suitable for transportation or further on-site treatment prior to ultimate disposal.

TABLE 1. Alternative Technologies Proposed for Destroying Chemical Weapons

HIGH TEMPERATURE (600-3,500 C)	**LOW TEMPERATURE** (0-200 C)
Molten Fe, Ni, or Cu Alloy	**Caustic Hydrolysis**
Molten Aluminum Alloy	**Water Hydrolysis**
Molten Zn, Tn, or Pb Alloy	**Electrochemical Oxidation with Ag**
Carbon Electrode Technology	Electrochemical Oxidation with Ce
Plasma Torch Technology	Enzyme Process
Induction Coupled Plasma Torch	Membrane Separation
Molten Salt Oxidation	X-ray Irradiation
Alkali Metal Reaction	Electron Beam Irradiation
Thermal Media Oxidation	**UV Irradiation and Oxidation**
Gas-Phase Reduction with Hydrogen	**Solvated Electron Technology**
Gasification	Microwave Irradiation
MEDIUM TEMPERATURE (200-600 C)	Plasma Corona
Hydrotreating	Neutralization and Encapsulation
Supercritical Water Oxidation	Anhydrous Liquid Halide
Wet Air Oxidation	Hydrogenation (Base Catalyst Decomposition)
Steam Reforming	Neutralization, Separation, and Incineration
Reforming in Water	Neutralization with Powdered Sodium
Catalytic Oxidation	Pretreatment, Neutralization, and Stabilization
Reaction Sulfur	Catalytic Neutralization Process
Sulfur and Lime Reactions	**Biodegradation**
Acid Oxidation	
Acid Roasting	

Moderate-temperature, high-pressure oxidation technologies continue to show some potential for both agent and energetics destruction; however, the high-pressure environment and corrosive nature of the reactants and products create materials selection challenges that will require an investment in research and pilot-plant development.

High-temperature and low-pressure oxidation technology alternatives, are in many ways similar to incineration, and also demonstrate the capability to efficiently destroy agent and energetics, and to decontaminate metal parts, in a single-step process. They also offer broad versatility in dealing with other types of wastes.

Pre-processing and post-processing operations must be used to support the basic reaction process for all alternative technologies as well as combustion reactions used in incineration. Often, critics of incineration focus on the primary or secondary combustion chamber as the critical unit process, while others choose to identify stack emissions, liquid discharges, or ash disposal as the critical unit processes (see Figure 1). Incineration and alternative technologies both require similar, and in many cases the same, unit processes that either precede or follow the main reaction chamber. Table 2 and 3 list examples of standard unit pre-processing and post-processing operations.

TABLE 2. Examples of Standard Unit Pre-processing Operations

Agent draining	Handling of energetics
Blending	Metal decontamination
Size reduction	Reagent handling
Mixing	Reagent feed control
Conveying	Air compression
Pumping	Steam generation
Munition handling	Agent feed control
Munition disassembly	Atomization of agent feed.

TABLE 3. Examples of Unit Post-processing Operations

Effluent tempering	Liquid filtration
Adjusting pH	Condensation
Agglomeration	Evaporation
Physical separation	Heat recovery
Acid gas removal	Liquid effluent handling
Removal of toxic organics	Solid effluent handling
Removal of toxic metals	Chemical or physical stabilization
Liquid recycle	Transportation
Gas filtration	Ultimate disposal or recycle.

To emphasize how unit process steps fit into the overall facility design, a conceptual layout of a standard Liquid Incineration facility is illustrated in *Figure* 1 and a Supercritical Water Oxidation facility is illustrated in *Figure* 2. The reaction chambers are shaded to emphasize the large number of pre-processing and post-processing unit operations that are required to support each technology.

To aid in the consideration of the range of possible gaseous, liquid, and solid effluents that may be produced by alternative technologies, a brief technical description of the technologies considered at the workshop follows. For a more detailed description, see Appendix A.

2. Low-Temperature, Low-Pressure Detoxification

2.1. CHEMICAL DETOXIFICATION

With chemical detoxification, agents are converted to products of reduced toxicity by chemical reaction with added reagents and solvents. The reactions are carried out at atmospheric pressure in the temperature range 20 to 100°C (68 to 212°F). Candidate reactions include hydrolysis in strong aqueous base; acid-catalyzed hydrolysis; reaction with ethanolamine; acid chlorinolysis; and reaction with chemical reducing agents.

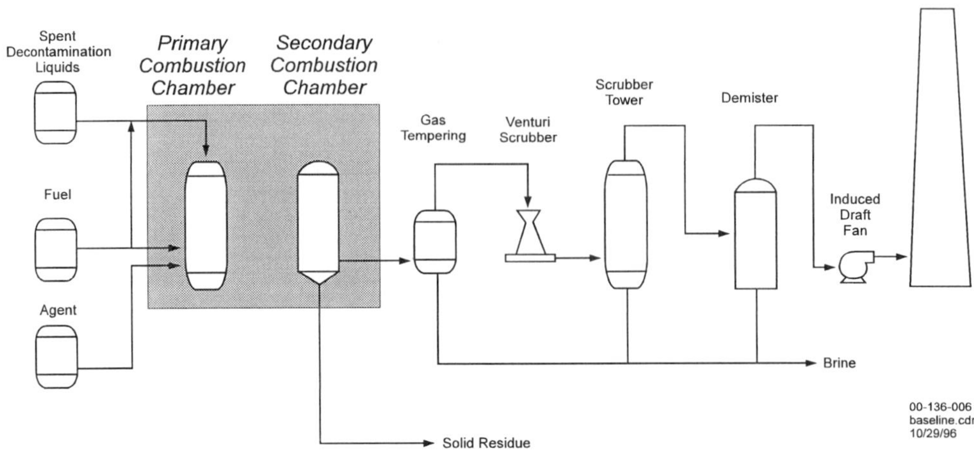

Figure 1. Liquid Incineration

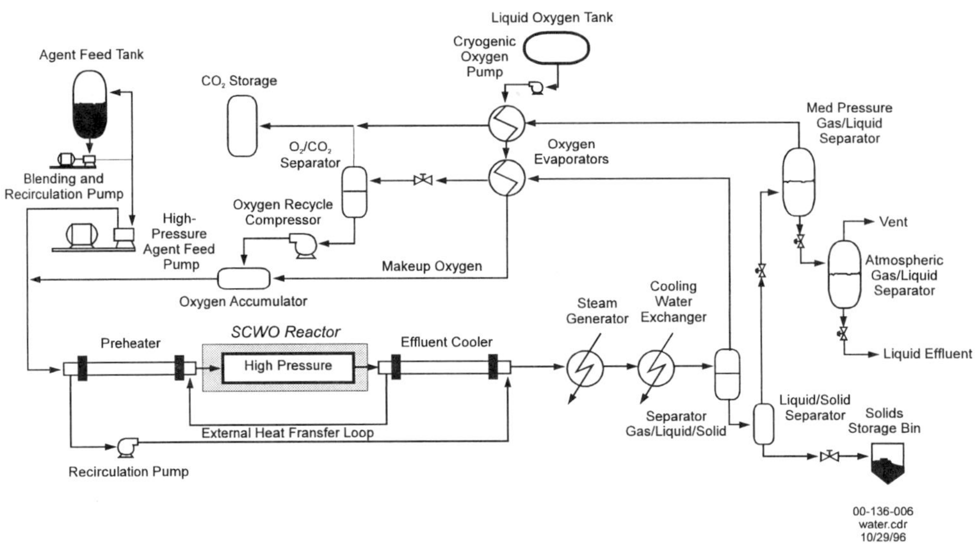

Figure 2. Supercritical Water Oxidation

2.2. BIODEGRADATION

Biodegradation involves the use of enzymes or cellular systems to degrade nerve agents or reaction products in dilute (approximately 10 percent) aqueous solutions.

2.3. SOLVATED ELECTRON CHEMISTRY

"Solvated electron solutions" are powerful reducing agents, capable of reacting with agent and other hazardous organic compounds to produce less hazardous organic species and inorganic salts to be disposed of in an environmentally safe fashion. These solutions are produced by dissolving such alkali or alkaline earth metals as sodium, potassium, lithium and calcium in anhydrous liquid ammonia.

2.4. ELECTROCHEMICAL OXIDATION

The electrochemical oxidation process operates below 212 °F (100°C) and ambient pressure using a nitric acid reaction medium. The process relies on the use of an electrically charged silver catalyst to generate highly reactive conditions for toxic compounds, oxidizing them to carbon dioxide (CO_2), water and inorganic compounds. Example by-products from this process include a gas stream containing nitrogen (N_2), hydrogen (H_2), nitrogen oxides (NO_x), carbon dioxide (CO_2), carbon monoxide (CO), a brine containing metallic salts, and waste acid.

3. Moderate-Temperature, High-Pressure Oxidation

3.1. WET AIR OXIDATION (WAO)

Wet Air Oxidation involves the oxidation of chemical substances in dilute aqueous medium at 200 to 335°C (392 to 635°F) and 20 to 136 atmospheres (294 to 2,000 pounds per square inch). Air is the source of the oxidizing agent.

3.2. SUPERCRITICAL WATER OXIDATION (SCWO)

Supercritical water oxidation reacts combustible materials using air or oxygen in water at temperatures and pressures above the critical point of water (374°C [705°F] and 218 atmospheres [3,205 psi]). Oxidation is accomplished under supercritical conditions to obtain excellent dense-phase mixing and high rates of heat transfer.

4. Moderate-Temperature, Low-Pressure Reactions

4.1. SULFUR PROCESS

The sulfur process reacts organic compounds with elemental sulfur, either in a vapor phase or liquid phase at temperatures ranging from 275-850°F (135-450 °C). Higher temperature may be present in the vapor phase. In the case of chemical warfare agents, reactions in molten sulfur or sulfur vapors are analogous to oxidative reactions; however, the reactions occur at lower temperatures and produce different by-products. Example by-products from the sulfur reactions include hydrogen sulfide gas (H_2S), sulfur dioxide (SO_2), carbon disulfide (CS_2), hydrogen chloride (HCl), nitrogen oxides (NO_x, depending on the presence of oxygen), phosphorus sulfides, sulfur chlorides and oxychloride, and a polymeric form of carbon-sulfur.

5. High-Temperature, Low-Pressure Reactions

5.1. STEAM REFORMING REACTOR

Steam reforming can be used to destroy chemical warfare agents by reacting them with superheated steam. A synthesis gas consisting of CO, CO_2, H_2, H_2O, and CH_4 is produced from these wastes. Steam reforming is mainly a two-stage operation, one stage where organics are vaporized by exposure to 316°C (600°F) to 593°C (1100°F) superheated steam, and a second stage, where the first stage offgas is mixed with 1093°C (2000°F), or greater, superheated steam to complete the reforming reactions. Steam reforming (a) is performed at atmospheric pressure, (b) significantly reduces waste volume, and (c) takes place in a low oxygen, reducing environment.

5.2. PLASMA TORCH REACTOR

A plasma torch design utilizes a refractory lined vessel heated by a plasma torch. A carrier gas flows through the torch (air, oxygen, nitrogen, etc.) and becomes a high temperature ionized plasma gas. Although centerline plasma temperatures may reach 12,000°C (21,632°F) or more, plasma surface and surrounding gas temperatures vary between 1500°C (2732°F) and 5300°C (9572°F). When air or oxygen is used as the plasma gas and secondary steam added as an oxidant and source of hydrogen, the reactor can produce a fuel or synthesis gas. The plasma reactor can process agent, metal, and other wastes.

5.3. HIGH-TEMPERATURE GAS-PHASE REDUCTION

The high-temperature (800-850 °C) gas-phase reduction process is destroys chemical warfare agents in an atmosphere of steam and excess hydrogen. The reduced chemical agents are converted to simpler molecules (CO, CO_2, HCl, NH_3, PH_3, H_2S) and a solid residue from metal or minerals that are introduced with the feed into the process. Gas removed from the process passes through a multiple stage scrubber system to remove water, mineral acids, and carbon dioxide (CO_2). The remaining synthesis gas is then compressed and stored in tanks until it can be tested and certified to be clean.

6. Effluents from Demilitarization Technologies

Based on the fundamental chemical reactions, the quantity of effluents from the various alternative technologies can be estimated. Table 4 presents estimates of the quantity of gaseous, liquid and solid effluents produced by each of several technologies for 3 chemical agents; VX, GB and HD (ref. 1).

The estimates of quantity of effluents in Tables 4 are theoretical and could be impacted greatly by actual operating procedures. Some of the operating parameters that could increase the quantity of effluent produced include: excess reactant chemicals, excess air or oxygen, excess diluents for pH control, excess scrubbing fluids, and excess quenching fluids.

7. Conclusions

Based on the studies presented at this Advanced Research Workshop, we concluded that there are specific research needs in the area of Effluents of Alternative Demilitarization Technologies. The workshop participants examined the current technologies and policies that were presented and developed the following seven recommendations:

- There is a pressing need to rigorously characterize the products of incomplete reaction (PIRs) in effluents of alternative demilitarization technologies.

- In the use of biodegradation technologies, it is especially important to characterize the by-products of reaction and their partitioning in bioprocessing effluent streams.

- Since heat and mass balances are generally not available for alternative technologies, there is an urgent need to develop, at least, mass balance data on all alternative demilitarization technologies including PIRs (trace elements) and by-products.

- The formation of reactive or toxic by-products (CO, H2S, H2, etc.) in alternative technology effluent streams, brings an urgent need to perform detailed process safety reviews and ultimately total quantitative risk assessments.

- Although real-time monitoring in gaseous effluent streams is relatively mature, there is a pressing need to develop continuous real-time monitoring for liquid effluent streams for process control and for controlling toxic substances.

- Depending on the technology, there remains a need to provide management and control of dioxins/furans (and their precursors) in all waste streams.

- For the introduction of new alternative technology to a site-specific application, methods need to be established for engaging the public and technical community in the process of technology selection, risk assessment, waste minimization, and reducing the potential environmental burden.

TABLE 4. Estimated Mass of Effluents for Technologies Destroying Chemical Agent[a]

Technology Agent Treated	HD			VX			GB		
Effluent Phase	Gas	Liquid	Solid	Gas	Liquid	Solid	Gas	Liquid	Solid
Molten Metal (oxygen)	1.19b	1.46	0.21	1.73b	1.23	0.23	1.0	2.0	1.4
Carbon Electrode (air)	1.6	1.7	1.3	3.7	2.6	0.7	2.5	1.7	1.2
Plasma Torch (air)	1.6	1.7	1.3	3.7	2.6	0.7	2.5	1.7	1.2
Induction Coupled Plasma (air)	1.6	1.7	1.3	3.7	2.6	0.7	2.5	1.7	1.2
Molten Salt Oxidation	16.7	1.6	1.4	24.7	4.9	0.2	17.5	2.2	3.0
Gas Phase Reduction with Hydrogen	0.77	2.22	...	0.74	0.46	0.55			
Hydrotreating	1.2	1.5	1.8	1.6	2.6	1.3	1.4	1.7	1.3
Supercritical Water Oxidation	1.4	23.1	...	2.3	25.5	...	1.6	22.1	...
Wet Air Oxidation	6.2	17.3	...	12.2	30.0	...	6.5	21.0	...
Steam Reforming	6.4	...	1.6	11.9	...	1.2	8.6	...	1.2
Reaction with Sulfur	0.0	3.3	0.5	0.7	0.7	1.3	0.7	2.0	0.8
Electrochemical Oxidation with Silver	4.11	13.84	6.15	6.72	23.7	8.63			
Water Hydrolysis	1.65	29.53	...						
Biodegradation of Hydrolysis Products	67.94	92.90	...						
Caustic Hydrolysis				1.73	11.04	...	...	3.9	...
Catalytic Oxidation	19.8	...	0.1	20.0	...	0.1	19.7	...	0.1
Thermal Media Oxidation	19.8	...	0.1	20.0	...	0.1	19.7	...	0.1
Gasification	1.6	1.6	1.4	3.5	2.8	0.7	2.5	1.9	1.2

a Pound of effluent per one pound of Agent destroyed

b Values with two significant figures after the decimal point are based on agent destruction data reported by the respective vendors.

8. Acknowledgments

I want to express a special acknowledgment to the authors who contributed technical papers for this book and to their countries and employers for supporting this effort. Our hosts, Prof. Katarina Demnerova and Prof. Rudolf Zahradnik from the Academy of Sciences of the Czech Republic provided excellent facilities and meeting support. The NATO Science Committee deserve special recognition; first, for arranging a planning meeting in Prague in June, 1995 where the idea for this workshop was formed, and second, for providing financial support and specific guidance through Mrs. Nancy Schulte, NATO Program Director. Finally, for editing, re-writing and preparing this document for publishing, I gratefully acknowledge the work and technical assistance provided by Heidi L. Wheeler, German Department, Johns Hopkins University and Dr. Darrell Wheeler, Philosophy Department, Towson University.

9. References.

1. F.W. Holm (1996, November) “Alternative Technologies for Destruction of the Chemical Weapons Stockpile,” Final Report for the U.S. Army, SAIC Delivery Order 136, Aberdeen Proving Grounds, MD.

U. S. CHEMICAL STOCKPILE DISPOSAL PROGRAMME: THE SEARCH FOR ALTERNATIVE TECHNOLOGIES[1]

RICHARD S. MAGEE
Center for Environmental Engineering and Science
New Jersey Institute of Technology
University Heights
138 Warren Street
Newark, New Jersey 07102

1. The Call for Disposal

The United States has maintained a stockpile of highly toxic chemical agents and munitions for more than half a century. Chemical agents are extremely hazardous, which is why they have been used in weapons. The manufacture of chemical agents and munitions and their subsequent stockpiling were undertaken in the belief that they had value as deterrents to the use of similar materials against U.S. forces. Today, other deterrents are considered more appropriate. In an attempt to avoid the worldwide risk posed by chemical warfare, the United States has entered into an agreement with many other nations to rid the world of all chemical weapons and munitions. Even apart from this agreement, the United States can no longer justify the continuing risk and expense of storing them. Consequently, there is ample incentive for the United States to dispose of its chemical agents and munitions as soon as this can be done safely.

2. Description of the Stockpile

2.1. AGENTS

The principal unitary chemical agents in the U.S. stockpile are the two nerve agents (GB and VX) and three related forms of blister, or mustard, agent (H, HD, and HT). These agents are stored in a variety of munitions and containers and exist largely as liquids. The stockpile consists of 30,600 tons of unitary agents (U.S. Army, 1996).

---

[1]Abstracted from the National Research Council report, Review and Evaluation of Alternative Chemical Disposal Technologies, Panel on Review and Evaluation of Alternative Chemical Disposal Technologies, National Academy Press, Washington, D.C. 1996; Board on Army Science and Technology

F.W. Holm (ed.), Effluents from Alternative Demilitarization Technologies, 1–12.

2.2. CONTAINERS AND MUNITIONS

Unitary chemical agents are stored in spray tanks, bulk storage (ton) containers,[2] and a variety of munitions including land mines, M55 rockets, bombs, and artillery and mortar projectiles. Some munitions contain no explosives or propellant, whereas others contain some combination of fuse, booster, burster, and propellant. These components are referred to collectively as "energetics." These energetics incorporate a variety of chemical compounds that must also be eliminated as part of the Army Chemical Stockpile Disposal Program (CSDP).

2.3. GEOGRAPHICAL DISTRIBUTION

The unitary chemical stockpile is located at eight continental U.S. storage sites (see Figure 1) and at Johnston Atoll in the Pacific Ocean about 700 miles southwest of Hawaii. Table 1 gives the composition of the stockpile at each continental U.S. site by type of container or munition and by type of agent.

3. Role of the National Research Council

In 1985, Public Law 99-145 mandated an "expedited" effort to dispose of M55 rockets containing unitary chemical warfare agents, which are particularly hazardous munitions during storage, because of the potential for self-ignition. This program soon expanded into the CSDP, whose mission was to eliminate the entire stockpile of unitary chemical weapons. The CSDP developed the current baseline incineration system. In 1992, after setting several intermediate goals and dates, Congress enacted Public Law 102-484, which directed the Army to dispose of the entire stockpile of unitary chemical warfare agents and munitions by December 31, 2004. Since 1987, the Committee on Review and Evaluation of the Army Chemical Stockpile Disposal Program (the Stockpile Committee) of the National Research Council (NRC) has overseen the Army's disposal program and has endorsed the baseline incineration process as an adequate technology for destroying the stockpile.

Growing public concerns about, and opposition to, incineration, coupled with the rising cost of the CSDP, has raised interest in alternatives. The Stockpile Committee, which has been following the state of alternative technologies, reviewed a NRC study of alternative technologies by a separate NRC committee and in 1994 recommended that the Army continue research on neutralization.

In the summer of 1995, the assistant secretary of the Army for research, development and acquisition informally explored the issue of examining alternative chemical disposal technologies with the Stockpile Committee. Following numerous discussions between the Army and the NRC, a decision was made to conduct a new NRC study to reexamine the status of a limited number of maturing alternative chemical disposal technologies (including the two neutralization-based processes on

---

[2]Although bulk containers are commonly referred to as "ton containers," they actually weigh 635.6 kg (1400 lb.) empty and contain an additional 681 to 726 kg (1500 to 1600 lb.) of agent. The total weight is approximately 1407 kg (3100 lb.) (U.S. Army, 1988).

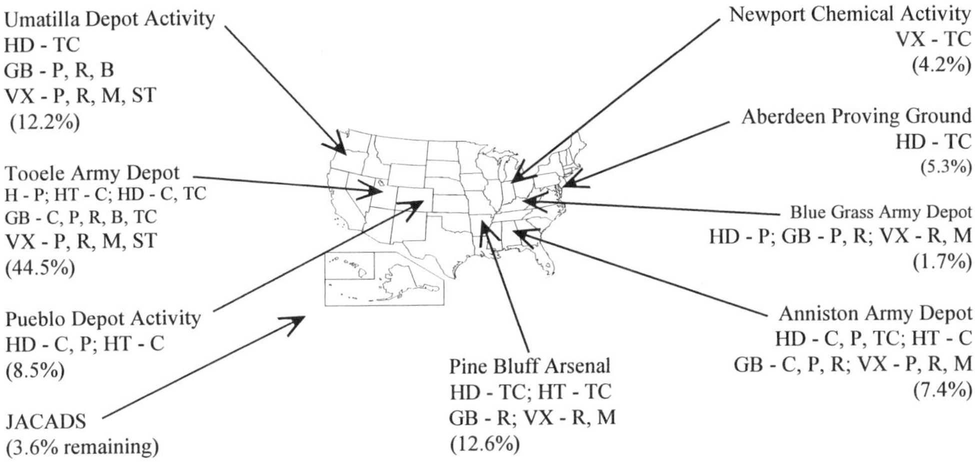

Figure 1. Types of Agents and munitions and percentage of total agent stockpile at each storage site. Derived from OTA, 1992; NRC, 1996.

TABLE 1. Chemical Munitions Stored in the Continental United States

Chemical Munitions (Agent)	APG	ANAD	BGAD	NECA	PBA	PUDA	TEAD[a]	UMDA
Mustard agent (H, HD, or HT)								
105-mm projectile (HD)		X				X		
155-mm projectile (H, HD)		X	X			X	X	
4.2-in. mortar (HD, HT)		X				X	X	
Ton container (HD)	X	X			X		X	X
Ton container (HT)					X			
Agent GB								
105-mm projectile		X						
155-mm projectile		X	X				X	
8-in. projectile		X	X				X	X
M55 rocket		X	X		X			X
500-lb bomb							X	X
750-lb bomb							X	X
Weteye bomb							X	X
Ton container		X[b]					X	
Agent VX								
155-mm projectile		X	X				X	X
8-in. projectile							X	X
M55 rocket		X	X		X		X	X
M23 land mine		X			X		X	X
Spray tank							X	X
Ton container				X			X	
Miscellaneous								
Ton containers (L)							X	
Ton containers (GA)							X	

[a]Small quantities of Lewisite and tabun (GA) are stored in ton containers at TEAD.
[b]Small quantities of agent drained as part of the Drill and Transfer System assessment for the M55 rockets.

NOTE: APG, Aberdeen Proving Ground, Maryland; ANAD, Anniston Army Depot, Alabama; BGAD, Blue Grass Army Depot, Kentucky; NECA, Newport Chemical Activity, Indiana; PBA, Pine Bluff Arsenal, Arkansas; PUDA, Pueblo Depot Activity, Colorado; TEAD, Tooele Army Depot, Utah; and UMDA, Umatilla Depot Activity, Oregon

SOURCE: Adapted from NRC, 1996 and U.S. Army, 1996.

which the Army was currently conducting research) for possible implementation at the two bulk-storage sites at Aberdeen Proving Ground, Maryland, and the Newport Chemical Activity, Indiana.

The NRC established the Panel on Review and Evaluation of Alternative Disposal Technologies (the AltTech Panel) to conduct the new study. The panel included six members of the Stockpile Committee, who have accumulated experience in dealing with the complex issues involved in monitoring the destruction of the unitary chemical agent stockpile, and eight new members, who possess specific expertise for thoroughly evaluating the alternative technologies.

4. Evaluation Factors

The AltTech Panel essentially adopted three of the four primary factors identified by the Stockpile Committee in the *Criteria Report Evaluation*: process efficacy, process safety, and schedule (NRC, 1995, pp. 14–19). The fourth factor, cost, was evaluated independently by the Defense Acquisition Board. In adopting these factors, the AltTech Panel modified the wording of the first two factors (modified portions are shown in italics):

1. *Process Efficacy.* Does the alternative agent-destruction process, when integrated with other necessary destruction system components, effectively *and reliably* meet agent-destruction requirements?

2. *Process Safety.* Is the alternative technology safe *and does it protect public health and the environment*? The criterion of "safe" adopted by the Stockpile Committee is minimization of total risk[3] to the public and to the environment (NRC, 1994).

3. *Schedule.* What are the impacts of implementation of an alternative technology on the schedule for stockpile destruction?

Each primary factor has several sub-factors, which may be interdependent. A negative judgment on a technology for a specific sub-factor need not imply a negative overall judgment for the primary factor. The sub-factors and their interdependencies are discussed below.

4.1. PROCESS EFFICACY

Process efficacy encompasses not only the capability of a technology to destroy the agent of interest, but also the status of the technology: its stage of maturation along a spectrum from laboratory scale to pilot-plant development and eventual full-scale operation. Process efficacy also includes whether the process can be controlled, whether it is reliable, and whether it meets applicable regulatory and treaty

---

[3] Total risk is the cumulative adverse consequences from all relevant risks—for example, storage, transport, and processing risks—over the full remaining duration of the stockpile's existence and the stockpile disposal program.

requirements. The AltTech Panel defined the following sub-factors under process efficacy:

- technology status
- capacity to detoxify agent
- satisfaction of treaty requirements
- satisfaction of environmental and other regulatory requirements
- management of process residuals[4]
- process stability, reliability, and robustness
- process monitoring
- natural resource requirements (e.g., energy)
- scale-up requirements
- applicability for treating other wastes

5. Process Safety

Process safety encompasses concerns about worker safety, community health risks, and environmental protection. Therefore, evaluating process safety includes assessing in-plant safety and health risks, risks to community safety and health, and risks to the environment. For each of these major risk categories, the evaluation should include the consequences of a release of chemical agent and of non-agent, toxic process residuals. Important contributing factors to the overall risk in each category include the risks from storing and handling agent in containers prior to processing, as well as the risk of releases from the destruction process itself.

The discussion below covers, in broad outline, the full range of risk factor evaluation and of risk assessment, preliminary and quantitative, that must be done in the course of developing an alternative technology from pilot-testing to the construction of a full-scale operational facility.

For this particular study, time constraints and the immaturity and status of design of the candidate technologies precluded making quantitative risk assessments. However, the panel was able to:

1. make a qualitative evaluation of whether each technology can be operated safely, given the current state of development (assuming adequate attention is paid to the intrinsic safety issues for each technology),
2. identify the intrinsic safety issues for each technology and evaluate the current treatment of these issues by the Technology Proponent Companies (TPCs), and
3. provide focus for a future comprehensive, quantitative risk assessment prior to implementation.

---

[4]A process residual is defined as any material remaining at the end of the process. Process residuals include not only all materials in gaseous, liquid, or solid waste streams (emissions, effluents, and wastes) but also materials that may be considered products or by-products because they can be used or have economic value. Process residuals include residual agent or other materials that were in the process feeds (water, chemicals, etc.), as well as materials produced during processing.

6. Schedule

To compare the effect of alternative technologies on the implementation schedule for stockpile destruction, the panel needed estimated schedules for each alternative technology at each potential site. These technology-specific schedules had to include time ranges for technology development, pilot-scale evaluation, and full-scale implementation and operation. The panel requested schedules indicating major milestones—and the assumptions made in estimating them—for (1) laboratory and bench-scale development, if applicable; (2) pilot plant design, construction, and, operation, with subsequent analysis of pilot-plant data; and (3) design of the full-scale plant, acquisition of equipment, and the construction, startup, operation, and decommissioning of the full-scale facility.

Public opposition, regulatory review, and permitting requirements can cause significant delays in the implementation schedule, but informed public acceptance and support can help to overcome regulatory or statutory hurdles. The actual time required to implement a system and eliminate the stockpile will not only affect compliance with the CWC but will also significantly affect the overall risk at each site, because storage risk depends on the duration of storage.

The panel met with members of the communities near the Newport and Aberdeen sites, with representatives of the Indiana and Maryland Citizen Advisory Commissions (CACs), and with state regulators to solicit information and learn how these groups see issues affecting the implementation of each alternative technology. In particular, regulators were asked to provide information on technology-specific permitting requirements. CACs and local communities were asked to discuss their specific concerns about the technologies selected for evaluation and their views on criteria that should be used in the evaluation.

7. Technology Comparisons

Table 2 summarizes basic engineering data for each of the evaluated alternatives, including general process description, operating conditions, and the fate of the elements from destroyed agent (that is, the form of the process residuals containing elements from the agent).

8. Summary of Key Comparative Differences

Table 3 summarizes the discussions of how the AltTech Panel evaluated each alternative technology with respect to 11 comparison criteria. Unless otherwise noted, a table entry applies to both HD at Aberdeen and VX at Newport. The table provides a quick overview of the panel's evaluations, with emphasis on the differences among them.

TABLE 2. Process Engineering Data for Alternative Technologies

Engineering Parameter	Catalytic Extraction Processing	Mediated Electrochemical Oxidation	Gas-Phase Chemical Reduction	Neutralization of HD (Configurations 1, 2, 3)	Neutralization of HD (Configuration 4)	Neutralization of VX
Process Description						
Medium of Treatment	molten iron or nickel	8 M nitric acid	gas phase H_2 and steam	hot water	hot water	33% aqueous NaOH
Batch or Continuous	continuous	semibatch	continuous	semibatch	semibatch	semibatch
Operating Conditions						
Process Temperature (°C)	1600	90 (max)	850	90	90	90
Process Pressure	<10 atm. at injection; 1–2 atm above bath	near atmospheric	near atmospheric (slight positive pressure)	near atmospheric	near atmospheric	near atmospheric
No. of Unit Operations	> 12	10	> 10	7 (config. 1) to 5 (config. 3) on site	2 on site	3 on site
Electrical Power (kW·h/1000 kg)	7,400 net (HD), 25,000 excluding cogeneration	72,600 (HD) 134,900 (VX)	1,020	99,000 (conf. 1) to 23,000 (config. 3)	13,600	8,000
Fate of Agent: Ultimate Form (kg/1,000kg agent)[a]						
Carbon in HD	705 kg CO from HD; 166 kg CO from CH_4 cofeed	1,100 kg CO_2	30 kg C soot; remainder CO_2[b]	163 kg biosolids 4,563 kg CO_2 67–77 kg organics	624 kg thiodiglycol 125 kg other hydro-lysis products	
Sulfur in HD	201 kg elemental S	900 kg Na_2SO_4	201 kg elemental S	799 kg Na_2SO_4 67–77 kg organics 163 kg biosolids	624 kg thiodiglycol	
Chlorine in HD	460 kg HCl	750 kg NaCl	460 kg HCl	747 kg NaCl	715 kg NaCl; remainder in other hydrolysis products	
Carbon in VX	1,152 kg CO from HD; 136 kg CO from CH_4 cofeed	1,900 kg CO_2	52 kg C soot, remainder CO_2			463 kg EMPA-Na 49 kg MPA-2Na 931 kg S/P/N organics
Phosphorus inVX	116 kg P in iron alloy	600 kg Na_3PO_4	326 kg H_3PO_3[c]			554 kg P-containing organics
Sulfur in VX	120 kg S in alloy	550 kg Na_2SO_4	127 kg elemental S			864 kg S-containing organics
Nitrogen in VX	52 kg N_2 in offgas	300 kg $NaNO_3$	56 kg N as N_2/NH_3			828 kg N-containing organics

[a] The elemental composition of 1,000 kg of HD is 302 kg of carbon, 50.3 kg of hydrogen, 201 kg of sulfur, and 446.5 kg of chlorine. The elemental composition of 1,000 kg of VX is 494 kg of carbon, 97 kg of hydrogen, 116 kg of phosphorus, 120 kg of oxygen, 120 kg of sulfur, and 52 kg of nitrogen.

[b] Total carbon for GPCR includes carbon from natural gas reformed to CO and H_2, as well as carbon from agent.

[c] Appearance of phosphorus as H_3PO_3 or its salts is hypothesized by the panel as the most likely product from the reactor, based on thermodynamics. The actual P-containing process residuals from GPCR have not been demonstrated.

TABLE 3. Summary of Comparison Criteria for VX at Newport and HD at Aberdeen

Comparison Criterion	Catalytic Extraction	Electrochemical Oxidation	Gas Phase Reduction	Neutralization
Process Performance and Engineering				
Technology status	VX and HD destruction demonstrated at bench scale. Entering commercial operation for low-level radioactive waste.	VX and HD destruction demonstrated at bench scale. No commercial or pilot-scale operation on other wastes.	VX and HD destruction demonstrated at laboratory scale. Applied commercially (full-scale) to chlorine-containing organics. For VX, phosphorus-containing products and subsequent scrubbing yet to be determined.	VX and HD destruction demonstrated at bench scale. For VX, low toxicity/burden of hydrolysate or treatability needs validation.
Stability, reliability, robustness	Proven components tightly integrated into a well-controlled process.	Easily controllable oxidation at very low agent concentrations.	Several ($\geq$ 10) proven unit operations that require tight integration. No strongly exothermic reactions.	Low temperature, low pressure semibatch process. Standard equipment.
Safety, Health, and Environment				
Safety interlocking	High degree of integrated process control and safety interlocks are required and have been developed.	Minimum interlocking required; reactions can be stopped easily by shutting off power.	High degree of integrated process control and safety interlocks are required; high-temperature hydrogen; temperature and pressure control are critical.	Because of interstage storage, minimal interlocking required.
Hazard inventory	Agent under pressure in delivery system. A high temperature, moderately high pressure process. High thermal mass. Combustible and reactive offgases. For HD, large volume of toxic by-product H_2S.	Large volume of reactive reagents (HNO_3, H_2O_2, NaOH).	High temperature agent and combustible gas. Difficulty of preventing buildup of hydrogen and containing agent within a building. For HD, large volume of toxic by-product H_2S.	Concentrated sodium hydroxide.
Test prior to release	Provision made for testing gases prior to combustion. Solids and liquids can be tested before shipment. Combustion gases released without analysis through stacks.	All aqueous and solid residual streams can be tested prior to release.Gases treated extensively prior to release.	All product streams can be stored and analyzed before release. Combustion gases released without analysis through stacks.	Main (aqueous) waste stream is tested.
Environmental burden	Relatively low because of high degree of recycling, especially if syngas-to-energy is considered recycling.	Released residuals are common gases or salts in their most stable forms.	Low. Sulfur recycled. HCl and/or NaCl in stable, diposable, or recyclable form. State and disposition of all secondary wastes must be defined.	For HD, aqueous discharge contains salts (NaCl, Na_2SO_2) Biomass. For VX, same as HD except aqueous discharge also has Na_3PO_4.

TABLE 3. Summary of Comparison Criteria for VX at Newport and HD at Aberdeen (continued)

Comparison Criterion	Catalytic Extraction	Electrochemical Oxidation	Gas Phase Reduction	Neutralization
Worker safety	Process is complex but well developed. Preliminary FMEA indicates process meets safety standards.	Low temperature and low pressure. Requires handling reactive chemicals.	Hazard analysis for chlorine wastes developed. Analysis of more complex recovery/scrubbing systems required for agent.	Low temperature and pressure. Mild caustic. HD configs. 1, 2 have H_2O_2.
Implementation Schedule				
Technical development	For HD, advanced. For VX, advanced but not as far as for HD.	Appears to be straightforward, but technology least developed of those evaluated and much engineering development remains to be accomplished.	Advanced. Well-developed process for destroying organic wastes. For VX, phosphorus-containing products need to be determined. Integration of phosphorus recovery into process not demonstrated.	HD: 60% design status. Ready for permit application. VX: 60% design status. 6 months to resolve toxicity/treatability of hydrolysate.
Processing schedule	Approximately one year to destroy agent at each site, after systemization. Operations at Newport will not begin until Aberdeen activities are completed.	Controlled by number of modules in facility.	For each site, approximately one year to construct and systematize, one year to destroy stockpile.	15 months to systemization. Less than one year to destroy stockpile, operating at full rate.
Permitting requirements	Current documentation adequate for timely review (unclear whether RCRA permit required).	Process novelty could lengthen permit review time.	Full-scale facilities permitted outside U.S. Permitting strategy submitted to panel.	Regulators indicate technology would take least time to permit.
Public acceptance	Attractive if seen as recycling. High temperature and pressure not attractive to public. Stack emissions may be a concern. BDAT designation for incinerable wastes may be positive (proven technology) or negative (versatile for other wastes).	Meets key preferences of public: low temperature, atmospheric pressure; closed loop.	Perceived as a closed loop system with provision for test before release. Low pressure but high temperature. Hydrogen may be perceived as a risk. Stack emissions may be a concern.	Concern about Army management of the technology. Low temperature and pressure. Closed loop batch process. Testing before release. CACs favor it. Agent-specific; not easily applied to other wastes.

9. Findings and Recommendations

The following findings and recommendations are based on the AltTech Panel's in-depth technical evaluation and assessment of five alternative technologies.

9.1. GENERAL FINDINGS

- All the technologies selected for the panel to review successfully demonstrated the ability to destroy agent at laboratory scale.

- As complete processing systems for chemical agent, all the technologies reviewed are of moderate to high complexity. Although components of each process are standard and proven, no alternative is an off-the-shelf solution as an agent-destruction process. Any one of them will require extensive design review, hazard and operability studies, materials selection, and related work as it moves through the piloting stage to full-scale demonstration and operation. During this necessary preparation for implementing an agent-destruction system, everyone involved should bear in mind that most failures in complex, engineered systems occur not during steady-state, normal operations but during transient conditions such as startup, shutdown, or operator responses to deviations from design conditions.

9.2. FINDINGS AND RECOMMENDATIONS FOR ABERDEEN AND NEWPORT SITES

9.2.1. Technology Selection

The panel's evaluation criteria favor technologies with the following characteristics:

- inherent process safety, which includes consideration of the required safeguards or engineering controls, the potential for process upsets, the requirements for process control interlocking, reliability, and the hazard inventory

- technical maturity, as shown by such factors as the scale of demonstrated ability to process agent and commercial industrial experience with the equipment, systems, and processes that would be required for an agent-destruction facility

- process simplicity, judged by such factors as the number of unit operations required and the ease of scale-up to a full-production facility

Based on these priorities, the panel reached consensus on the following findings and recommendations on alternative technologies to be pilot-tested for agent destruction at the Aberdeen and Newport sites.

9.2.2. HD at Aberdeen

Finding. Aqueous neutralization of the chemical agent HD followed by biodegradation of the hydrolysate surpasses the priority criteria listed above. This technology has the following advantages:

- Among the alternatives reviewed, it has the largest-scale successful demonstration with agent.
- The equipment required has been proven through extensive use in industry for processes similar to those planned for use in agent destruction.
- The principal unit operations are independent batch processes that do not require elaborate safety interlocking.
- Because the process involves batch processing of liquids, hold-and-test analyses to determine batch composition can be readily performed at several points in the process.
- The process is performed at low temperature and near-atmospheric pressure; the hazard inventory in general is low.
- The selection of materials of construction appears to be straightforward.
- No step in the process involves combustion; therefore no combustion products are emitted.

Recommendation. The Army should demonstrate the neutralization of HD at Aberdeen on a pilot scale.

- The AltTech Panel recommends biodegradation of hydrolysate from HD at an off-site treatment, storage, and disposal facility as the most attractive neutralization configuration. Of the four neutralization configurations, this one is the most reliable and robust; has little potential for process upsets; makes more use of existing facilities and trained staff, rather than requiring new facilities and newly trained staff; should be most rapidly permittable; should have the shortest implementation schedule; and should be the quickest and easiest to decommission.
- The second-best configuration is neutralization with biodegradation on site, followed by disposal of the aqueous effluent through a FOTW. If this option is selected, the panel recommends separating the volatile organic compounds (VOCs) prior to biodegradation, followed by off-site treatment and disposal of the VOCs.

9.2.3. VX at Newport

Finding. Neutralization of chemical agent VX with sodium hydroxide solution destroys agent effectively and substantially lowers the toxicity of the process stream. With respect to the priority criteria listed under "Technology Selection," this technology followed by off-site treatment and disposal of the hydrolysate has the same relative advantages as neutralization of HD. One difference, however, is the uncertainty about the appropriate disposal method for VX hydrolysate.

Although biodegradation of oxidized VX hydrolysate has been demonstrated in the laboratory, as of May 1996 limited treatability studies have not demonstrated biodegradation at a treatment, storage, and disposal facility (TSDF), even though a TSDF has disposed of VX hydrolysate from bench-scale testing within its permit requirements. It is possible, although not yet established by adequate testing, that the hydrolysate has sufficiently low toxicity associated with its organic products that complete biodegradation prior to discharge may not be necessary. Furthermore, treatment of VX hydrolysate by existing commercial TSDF processes other than biodegradation is likely to be possible. Therefore, any treatment at a TSDF, whether by biodegradation or another proven and tested process, that results in appropriately low toxicity and low environmental burden in the discharge from the TSDF is a suitable disposal option for VX hydrolysate. As an on-site option for the disposal of hydrolysate, the panel believes that existing, commercially proven processes other than biodegradation could be used. The residual concentrations of agent allowable under chemical-weapon treaty negotiations are likely to be less stringent than the concentrations required by the environmental permits for the destruction and downstream disposal facilities.

Recommendation A. The Army should pilot-test VX neutralization followed by off-site treatment of the hydrolysate at a permitted TSDF for potential use at the Newport site, only if the effluent discharged from the TSDF has been shown to have acceptably low toxicity and results in minimal environmental burden.

Recommendation B. If on-site disposal of VX hydrolysate is preferred to shipping it off site for TSDF treatment, existing commercial processes other than biodegradation should be considered. The panel does not recommend on-site biodegradation because of the need for co-feeding a substantial amount of carbon substrate and because of limited success to date in testing on-site biodegradation.

10. References

1. NRC (1994) Recommendations for the Disposal of Chemical Agents and Munitions, *National Research Council -- Committee on the Review and Evaluation of the Army Chemical Stockpile Disposal Program*, Washington, D.C.: National Academy Press.
2. NRC (1995) Evaluation of the Army's Draft Assessment Criteria to Aid in the Selection of alternative Technologies for Chemical demilitarization, *National Research Council -- Committee on Review and Evaluation of the Army Chemical Stockpile Disposal Program,* Washington, D.C.: National Academy Press.
3. NRC (1996) Review of Systemization of the Tooele Chemical Agent disposal Facility, *National Research Council -- Committee on Review and Evaluation of the Army chemical Stockpile Disposal Program*, Washington, D.C.: National Academy Press.
4. U.S. Army (1988) Chemical Stockpile Disposal Program Final Programmatic Environmental Impact Statement (PEIS), Aberdeen Proving Ground, Maryland: U.S. Army Program Manager for Chemical Demilitarization.
5. U.S. Army (1996) U.S. Chemical Weapons Stockpile Information Declassified, News Release, Office of Assistant Secretary of Defense (Public Affairs). Washington, D.C., January 22, 1996.
6. OTA (Office of Technology Assessment) U.S. Congress (1992) *Disposal of Chemical Weapons: An Analysis of Alternatives to Incineration,* Washington, D.C.: U.S. Government Printing Office.

MOBILE DEMILITARIZATION SYSTEM TREATMENT PROCESSES AND EFFLUENTS

EDMUND W. LIBBY
Project Manager for Non-Stockpile Chemical Materiel
Office of the Program Manager for Chemical Demilitarization
Aberdeen Proving Ground, Maryland 21010-5401

MARK D. CHATFIELD
Science Applications International Corporation
2109 Emmorton Park Road
Edgewood, Maryland 21010-1052

1. Introduction

The Program Manager for Non-Stockpile Chemical Material (PMNSCM) is responsible for disposing of all chemical warfare materiel (CWM) not specifically included in the U.S. unitary chemical stockpile. Included in this mission is the effective treatment and disposal of recovered chemical warfare materiel (RCWM), such as items recovered during range clearing operations and from CWM burial sites. RCWM consists of four distinct categories: chemical agent identification sets (CAIS), non-explosively configured munitions, explosively configured munitions, and bulk containers. It is anticipated that these items may be recovered at any or all of 97 locations in 38 states, the U.S. Virgin Islands, and the District of Columbia.

Disposal of RCWM poses many unique challenges. Unlike stockpiled munitions, RCWM are typically found in deteriorated condition. Often the munition and agent type cannot be identified by physical inspection of the item. The presence and status of explosives can also be difficult to discern. Further complicating the situation, corrosion of the metal munition body can create "leakers" which must be overpacked. The U.S. Army has developed a variety of non-invasive technologies that overcome most of the difficulties associated with initial munition assessment. However, RCWM in such poor physical condition are clearly not amenable to high speed demilitarization processes used for stockpiled weapons. More appropriate treatment technologies required development by the PMNSCM.

Taking into account the wide geographic dispersion of RCWM in the U.S., the PMNSCM began development of transportable chemical treatment systems in 1993. At that time, transporting the treatment system to the RCWM site was judged more effective than transporting RCWM from many different sites across State borders for ultimate disposal at one or more fixed facilities. Further consideration of the technical challenges and economics of operating these transportable treatment systems at remote sites led to the decision to dispose of process waste streams off-site. Based on the

F.W. Holm (ed.), Effluents from Alternative Demilitarization Technologies, 13–25.

knowledge that comparable classes of compounds found in industrial wastes are routinely processed at commercial treatment, storage and disposal facilities (TSDFs) in the U.S., the PMNSCM chose this approach as the post treatment step for process wastes generated by the first generation transportable treatment systems. In most cases, waste disposal at U.S. TSDFs employs incineration as the primary treatment step. In response to public concerns over incineration, the PMNSCM continues to evaluate alternatives to this post-treatment approach.

The PMNSCM is currently preparing to test two transportable systems designed to assess, treat and dispose of chemical agents and militarized industrial chemicals. The Rapid Response System (RRS) and the Munitions Management Device, Version 1 (MMD-1) employ chemical neutralization to dispose of CAIS and non-explosively configured munitions, respectively. Each system is designed to ensure environmental protection and the safety of workers and the public. The chemical neutralization processes achieve process simplicity for field use, result in marked reduction in agent characteristics, and generate waste streams that can be transported on public roads and processed at commercial TSDFs. The treatment processes and operational approaches of these systems are presented in greater detail in the following sections to illustrate the U.S. Army's current approach to RCWM disposal.

2. Rapid Response System

Historically, CAIS were used to train soldiers and sailors to detect and identify the threat of chemical materiels during warfare. They were produced in large quantities (approximately 110,000 sets) and various configurations from about 1928 until 1969. They are currently being recovered from burial sites in small quantities of boxes, metal shipping containers known as pigs, and as loose glass ampoules and bottles. The CAIS ampoules contain liquids, including chemical agents [neat sulfur mustard, 5% sulfur mustard (S-mustard) or Lewisite (L) in chloroform, 10% N-mustard in chloroform] and industrial chemicals (phosgene, cyanogen chloride, 50% chloropicrin in chloroform, and/or GA-simulant). CAIS bottles contain either liquids (neat S-mustard) or solids (S-mustard, N-mustard, L, or chloropicrin adsorbed on charcoal; triphosgene; chloroacetophenone; and Adamsite).

The RRS treatment process employs chemical neutralization to detoxify S-mustards, N- mustards and L. CAIS disposal is conducted within a glove box under engineering control in the RRS process trailer, as shown in Figure 1. CAIS containing chemical agents are segregated from those containing industrial compounds by inspection of original packing labels on the items or non-invasive Raman spectroscopic methods. CAIS ampoules and bottles containing industrial compounds are packaged for shipment to a commercial TSDF. CAIS ampoules and bottles containing agent are crushed inside a closed stainless steel reaction vessel in the presence of excess oxidizer and solvent. After the reaction is complete, the detoxified liquid waste and glass fragments are transferred to a liquid waste drum. Uncontaminated solid wastes from unpacking operations are transferred to a separate solid waste drum. Solid and liquid wastes are sampled and analyzed for residual agent and Resource Conservation and Recovery Act (RCRA) permitted constituents prior to shipment to a commercial TSDF.

Table 1 presents the RRS treatment processes. The BLUE process detoxifies neat S-mustards with the oxidizer 1,3-dichloro-5,5-dimethylhydantoin (DCDMH) in a solvent composed primarily of chloroform and tertiary butyl alcohol. In addition to strong oxidizing potential towards chemical agents, DCDMH was selected for RRS neutralization processes because of its relatively low reactivity towards chloroform. The RED process employs the same oxidizer/solvent system to treat dilute solutions of S- mustard, N-mustard, and L. The CHARCOAL process treats S-mustard and N-mustard adsorbed onto charcoal with DCDMH in pure chloroform solvent, and L on charcoal with same oxidizer/solvent system used for the RED and BLUE processes.

TABLE 1. RRS Treatment Processes

CAIS Materiel	Treatment Reagents	Neutralent Waste Stream Code
Neat S-mustard	DCDMH (in 48.5:48.5:3% volume mixture of chloroform, t-butyl alcohol, and H_2O)	BLUE
N-mustard (HN-1), S-mustard, and L in chloroform solution	DCDMH (in 48.5:48.5:3% volume mixture of chloroform, t-butyl alcohol, and H_2O	RED
N-mustard and S-mustard adsorbed onto charcoal	DCDMH (in chloroform)	CHARCOAL
L adsorbed onto charcoal	DCDMH (in 48.5:48.5:3% volume mixture of chloroform, t-butyl alcohol, and H_2O)	CHARCOAL "L"

Figure 2 illustrates the RED process chemistry for detoxification of dilute S-mustard in chloroform. The two primary reaction pathways lead to a variety of products including organic sulfoxides, sulfides and sulfones produced by chlorination/oxidation of S-mustard, chlorinated ethanes resulting from chlorination/fragmentation of S-mustard, and DCDMH-derived compounds. The BLUE process chemistry for detoxification of neat sulfur mustard is similar, but with a different product distribution, as shown in Figure 3.

Studies of the chemistries of the BLUE, RED and CHARCOAL processes have shown that the primary detoxification reactions are fast, with half-lives on the order of 30 seconds or less in solution. Residual agent concentrations in solution are reduced below the process goal of 50 parts-per million (ppm) within minutes of initiating the BLUE and RED processes. Due to the adsorption of agent in the micropores of charcoal, reduction of agent concentration below the process goal takes longer for the CHARCOAL process, on the order of thirty minutes. Secondary reactions between oxidizer and products or solvents also occur in the BLUE, RED, and CHARCOAL processes, but at slower rates. The three processes are well controlled, with low peak reaction temperatures and pressures, as shown in table 2.

TABLE 2. Reaction Parameters for the RRS Chemistry Processes

Reaction Parameter	BLUE Process	RED Process	CHARCOAL Process
Peak Temperature (°C)	~ 100	25	35
Peak Pressure (psig)	10	5	2.5

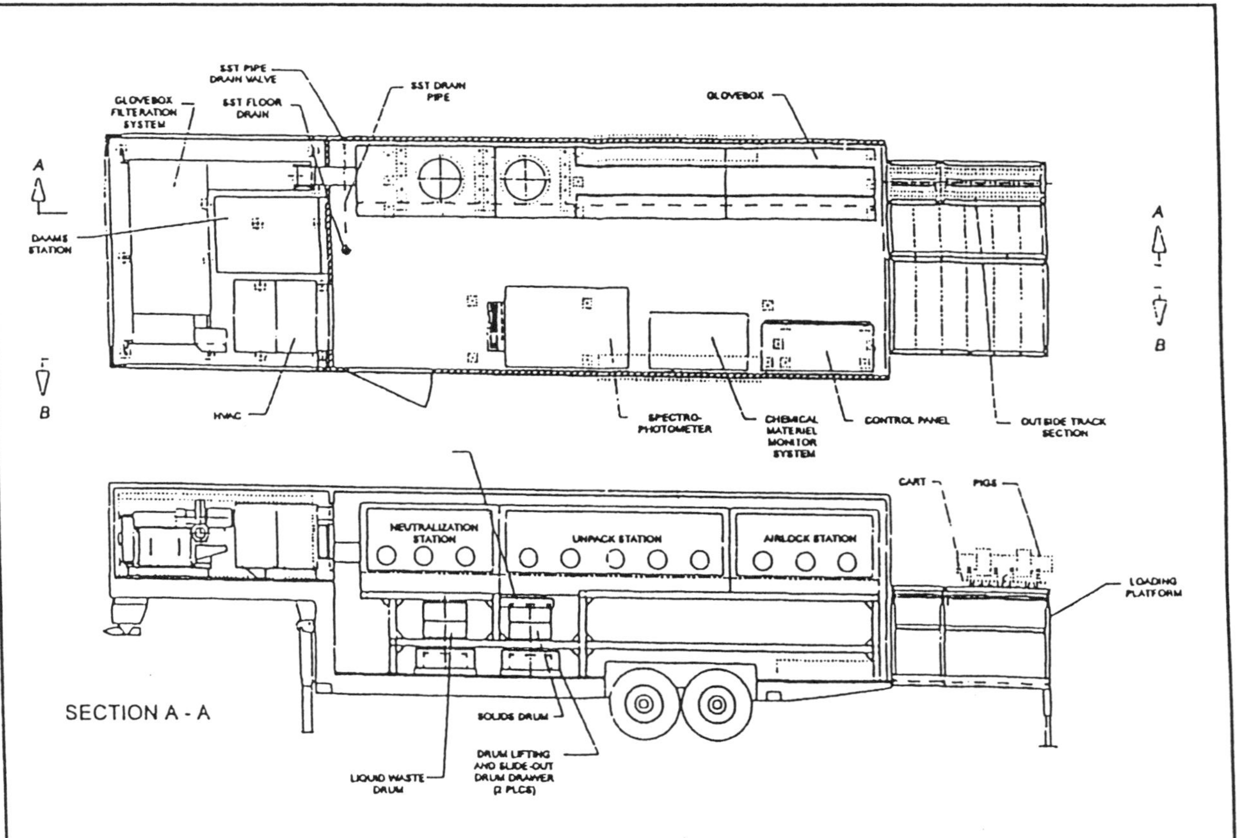

Figure 1. Rapid Response System

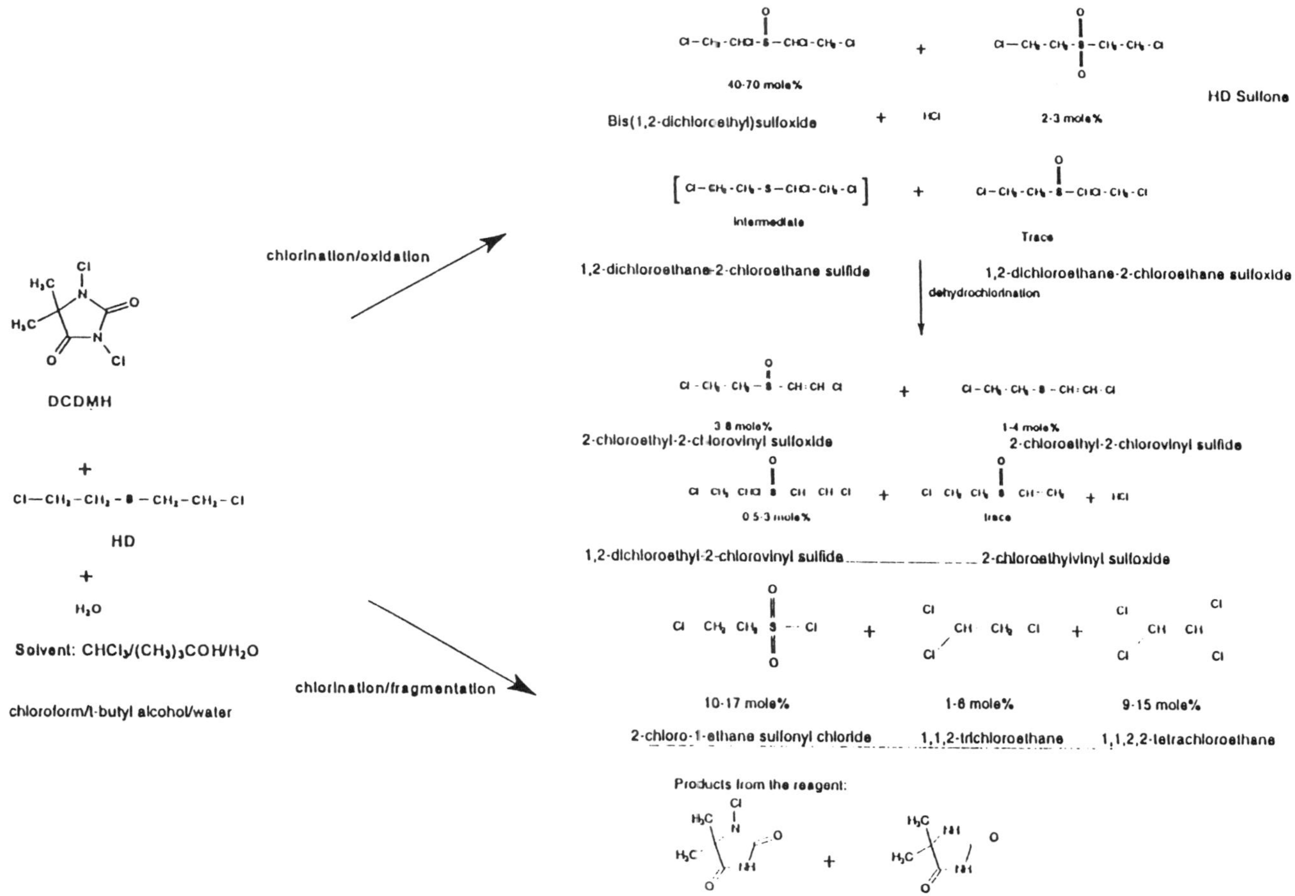

Note: Mole% represents conversion of HD to the product shown.

Figure 2. RRS RED Process Chemistry

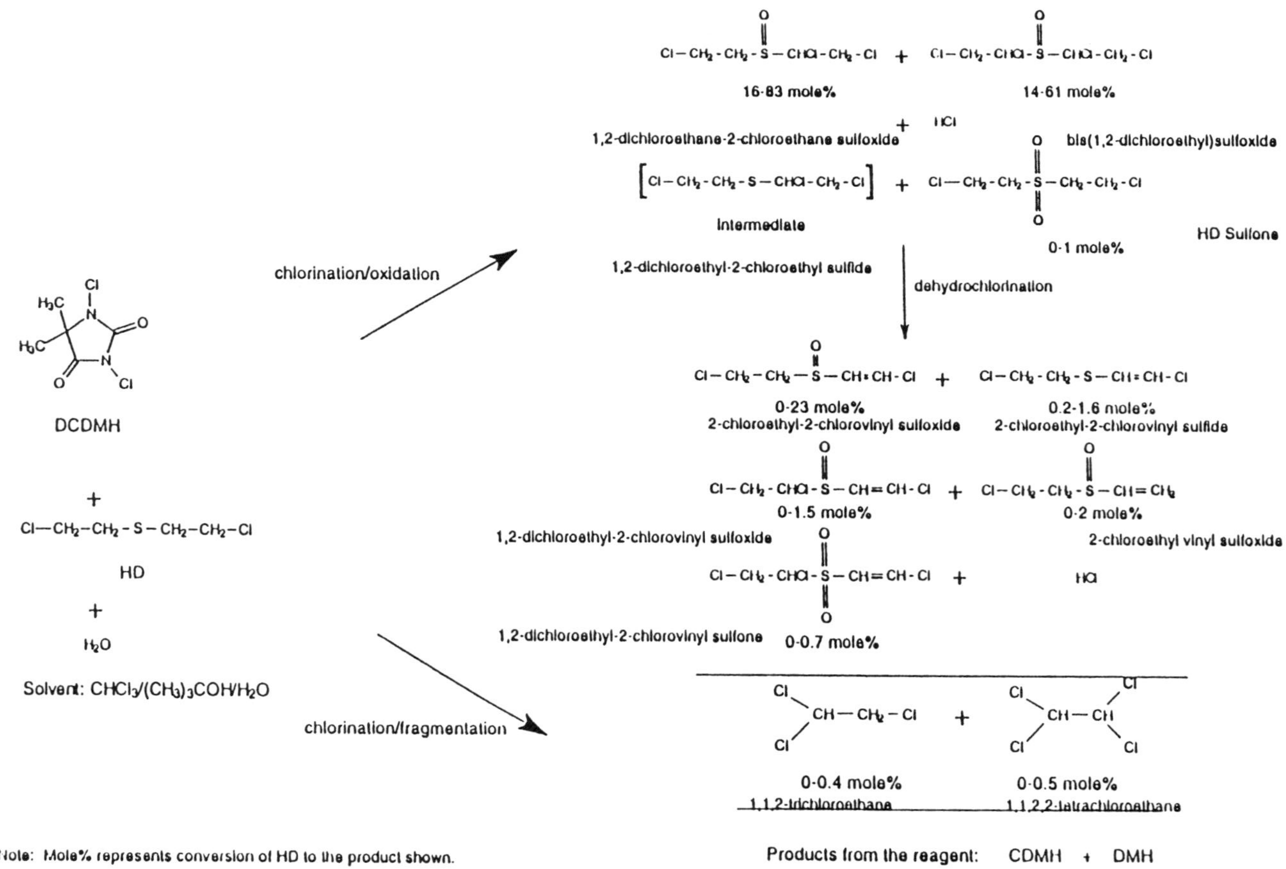

Note: Mole% represents conversion of HD to the product shown.

Figure 3. RRS BLUE Process Chemistry

Waste stream constituents for the BLUE, RED and CHARCOAL processes are presented in table 3. Liquid waste constituents include solvent, residual DCDMH oxidizer, and DCDMH-derived byproducts. Residual DCDMH is an inherent safety factor of the process, preventing the presence of unacceptably high concentrations of residual chemical agent in the waste. Reaction products from the neutralization of S-mustard, N-mustard and L include chlorinated sulfoxides and chlorinated sulfones, bis-(2-chloroethyl) amine, and chlorovinylarsonic acid, respectively. Byproducts of reactions between the oxidizer and solvent include a variety of chlorinated ethanes, butanes, and aldehydes. The chemical composition of these liquid waste streams suggests that worker exposure risk is dominated by the toxic effects of the solvent and residual oxidizer. PMNSCM conducted dermal irritation studies for the BLUE, RED and CHARCOAL processes to verify this hypothesis and comply with U.S. Department of Transportation (DOT) Hazardous Material Regulations (HMR) governing interstate transportation of hazardous materials.

TABLE 3. Rapid Response System Waste Stream Constituents

Waste Stream	Waste Stream Constituents
BLUE process neutralent [neat S-mustard (DCDMH)]	Chloroform, t-butyl alcohol, chloro-5,5-dimethylhydantoin, 5,5-dimethylhydantoin, Chlorinated sulfoxides (diethyl and ethyl vinyl), Chlorinated sulfones (diethyl and ethyl vinyl), 1,1,2-Trichloroethane, Tetrachloroethane, Chlorobutanes, sulfur mustard (< 50 ppm), water, glass and plastic
RED process neutralent [N-mustard, S-mustard, Lewisite/chloroform (DCDMH)]	Chloroform, t-Butyl alcohol, Chlorobutanes, 1,3-Dichloro-5,5-dimethylhydantoin, 5,5-dimethylhydantoin, Chlorinated sulfoxides (diethyl and ethyl vinyl), Chlorinated sulfones (diethyl and ethyl vinyl), bis-(2-chloroethyl)amine, Chlorovinylarsonic acid, 1,1,2-Trichloroethane, Tetrachloroethane, Acetaldehyde and chloroacetaldehyde, nitrogen mustard (< 50 ppm), sulfur mustard (< 50 ppm), Lewisite (< 50 ppm), water, insolubles, glass and plastic
CHARCOAL/CHARCOAL "L" Process Neutralent [N-Mustard, S-Mustard, Lewisite/charcoal (DCDMH)]	Chloroform, t-butyl alcohol, 1,3-Dichloro-5,5-dimethylhydantoin, Chloro-5,5-dimethyhydantoin, 5,5-dimethylhydantoin, Polychlorinated diethyl sulfides and polychlorinated ethyl vinyl sulfide, Chlorinated sulfoxides (diethyl and ethyl vinyl), Chlorinated sulfones (diethyl and ethyl vinyl), 1,1,2-Trichloroethane, Tetrachloroethane, Dichloroethane, Pentachloroethane, Hexachloroethane, Trichloroethylene, Chloral hydrate (or chloral), bis-(2 chloroethyl)amine, N-Chloropolychloroethylamines, Chlorovinylarsonic acid, Chlorobutanes and chlorobutenes, nitrogen mustard (< 50 ppm), sulfur mustard (< 50 ppm), Lewisite (< 50 ppm), water, charcoal, and glass
Miscellaneous Dunnage	Absorbent material, Metal PIG parts, Cardboard, Glass, Rags , Agent products, Agent/industrial chemical products, 5% sodium hypochlorite, DCDMH/chloroform/t-butyl alcohol/water

The dermal toxicity evaluations of neutralized CAIS waste streams were conducted in accordance with the Organization for Economic Cooperation and Development (OECD) guidelines [1]. The study was designed to determine whether the toxicity characteristics of the waste streams are suitable for designation with transportation packing groups stipulated for HMR Class 6 (Division 6.1) poisonous materials or HMR Class 8 corrosive materials. Both short-term acute dermal toxicity tests specific for Class 6 materials and dermal irritation studies specific for Class 8 materials were conducted. Quantitative analysis of skin irritation response data for both studies indicated that skin irritant effects from the BLUE, RED, and CHARCOAL processes are either comparable to or lower than those observed with the oxidant solvent system. These results verified that BLUE, RED, and CHARCOAL waste streams may be transported overland to a commercial TSDF per the shipping designations established for the oxidants and solvents.

3. Munitions Management Device Version 1

The MMD-1 is a transportable treatment system designed to provide onsite characterization and disposal of non-explosively configured RCWM in a safe, secure, and environmentally acceptable manner. The MMD-1 process trailer is shown in Figure 4. The MMD-1 accesses, removes and chemically neutralizes S-mustard, sarin, VX, and phosgene from artillery projectiles, mortar rounds, bomblets from cluster bombs, and bombs (up to 500 pounds in weight) that have no fuze or burster. Small containers of chemical materiel may also be treated in the MMD-1. The process also chemically decontaminates the munition body or container to acceptable levels for transportation and permanent disposal.

Prior to treatment, each chemical munition is characterized by noninvasive techniques such as radiography and neutron activation analysis. The munition is transferred to the munitions treatment vessel (MTV) within the MMD-1 where it is washed and then inspected via closed circuit television. The munition cavity is accessed under engineering controls by cutting a hole in the munition body. Chemical vapors are sampled and analyzed using a GC/MS/FTIR to confirm the identification of the munition contents. A large excess of agent-specific neutralization mixture is added to the MTV and recirculated for a preset time. The resulting neutralent is then pumped to liquid reactor vessel for further processing or to a waste holding tank. Gases such as phosgene are processed with a gas reactor. Following treatment, the agent cavity of munition body is cleaned with a high pressure water spray to remove solid materials. The final detoxified liquid and solid wastes are then transferred to RCRA permitted and DOT approved containers. As with the RRS, the MMD-1 wastes are sampled and analyzed for residual agent levels and RCRA permitted constituents prior to shipment to a commercial TSDF.

The neutralization mixtures used in the MMD-1 are based on reaction of chemical agents with Monoethanolamine (MEA) and water or aqueous caustic as shown in table 4. Phosgene is reacted with aqueous sodium hydroxide to form simple inorganic salts.

Reaction parameters are presented in table 5. Agent detoxification is rapid for phosgene and slower for the chemical agents, requiring up to 4 hours for sulfur mustard. Reaction temperatures are mild, with peak temperatures under 200 ^{o}F. A typical profile of reaction temperature and agent concentration for detoxification of HD with MEA/water at laboratory scale is shown in Figure 5 [2]. The bulk of the agent is destroyed within the first thirty minutes of reaction, with a corresponding peak in reaction temperature which slowly decays over the course of several hours as the residual sulfur mustard continues to be destroyed to levels below 1 ppm.

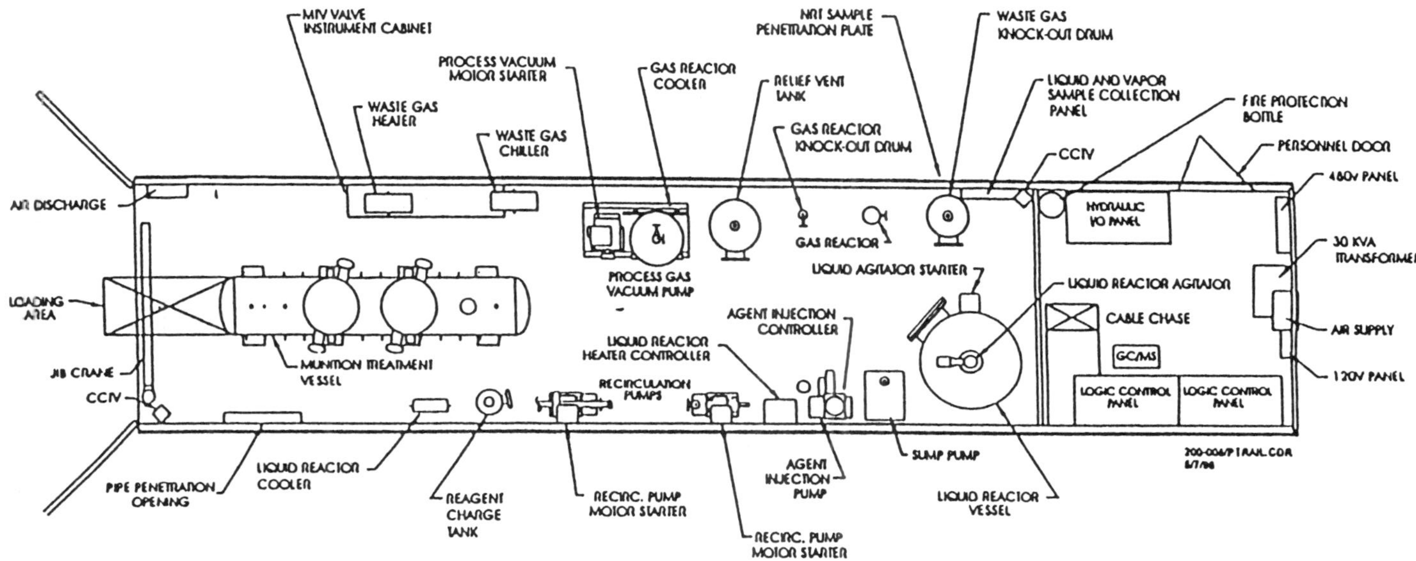

Figure 4. MMD-1 Process Trailer

TABLE 4. MMD-1 Treatment Processes

Chemical Agent/ Industrial Chemical	Reagent
S-Mustard	Monoethanolamine (MEA) and water
Sarin	MEA and water
VX	MEA and aqueous Sodium Hydroxide
Phosgene	Aqueous Sodium hydroxide

TABLE 5. Reaction Parameters for the MMD-1 Chemistry Processes

Reaction Parameter	S-Mustard	Sarin	VX	Phosgene
Starting Temperature (°F)	122	76-90	113-125	< 93 (liquid)
Peak Temperature (°F)	175	109-126	Continuous heating required	Rise of 3 – 73 °F is predicted
Reaction time for 99.98% destruction	< 4 hours at 122°F	< 3 hours at 76-126 °F	< 2 hours at 113-125 °F	rapid

Similar to the phase 1 Russian process, PMNSCM uses MEA and water to detoxify agent GB. A 45:55 volume ratio of MEA to water is used in the MMD-1 because it is effective and avoids formation of undesirable viscous product solutions, which occurs at lower concentrations of water. Two primary organic products, isopropyl methylphosphonic acid (IMPA) and 2-aminoethyl isopropyl methyl phosphonate (GB-MEA), are formed along with inorganic fluoride (HF) [3]. The two acids form soluble salts with MEA, as shown in figure 6. Reaction of sulfur mustard with MEA/water and VX with MEA/NaOH/water also produce a variety of reaction products, as listed in table 6. Residual agent is expected at concentrations in the low part per billion range. The primary constituents of the liquid wastes for HD, GB and VX neutralents are the excess neutralizing reagents. Similar to the RRS, the chemical composition of these MMD-1 liquid waste streams suggests that worker exposure risk is dominated by the toxic effects of the excess neutralizing reagents.

Dermal toxicity evaluations of neutralized MMD-1 waste streams were conducted in accordance with the OECD guidelines to determine whether the toxicity characteristics of the waste streams are suitable for designation with transportation packing groups stipulated for HMR Class 6 (Division 6.1) poisonous materials or HMR Class 8 corrosive materials [4]. Quantitative analysis of skin irritation response data for short-term acute dermal toxicity tests and dermal irritation studies indicated that skin irritant effects from the HD, GB, and VX neutralents are either comparable to or lower than that observed from the MEA/water and MEA/caustic reagents.

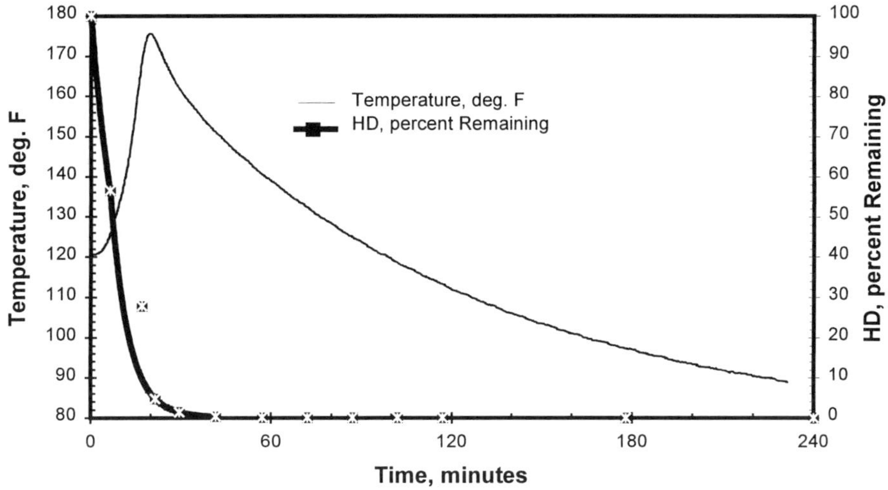

Figure 5. Temperature and Concentration Profiles for Detoxification of HD with MEA/Water

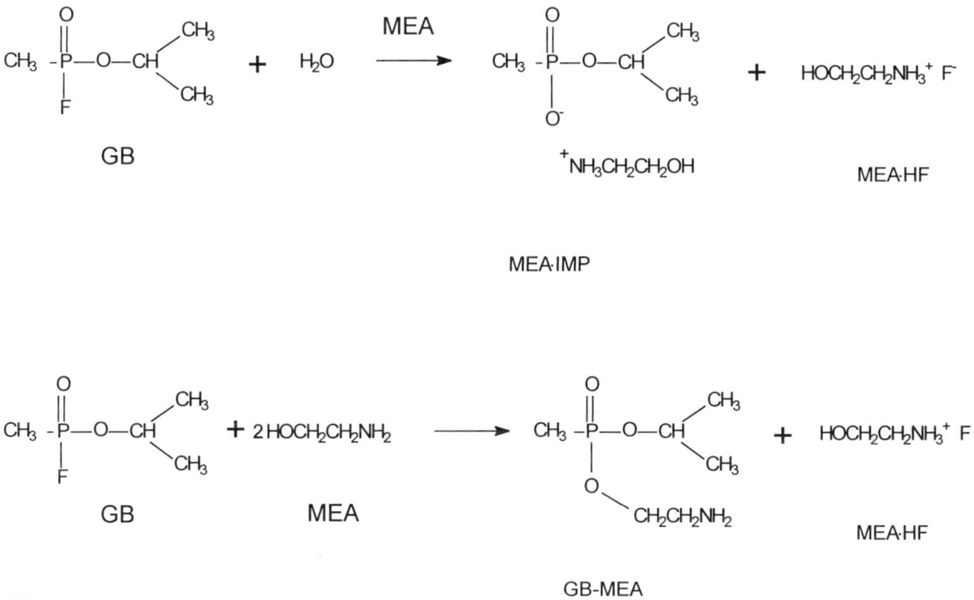

Figure 6. Reaction of GB with MEA/Water

TABLE 6. MMD-1 Waste Stream Constituents

Waste Stream	Waste Stream Constituents
S-Mustard Neutralent	MEA, water, 2-hydroxyethylammonium chloride, N-(2-hydroxyethyl)-thiomorpholine, bis-[(2-hydroxyethylamino)ethyl]sulfide and other organic sulfides, 1,4-dithiane, vinyl chloride, tetrachloroethylene, trichloroethylene, selenium, S-mustard (< 50ppb), 1,1-dichloroethylene, chloroform, methyl ethyl ketone, and trace metals
Sarin Neutralent	MEA, water, 2-hydroxyethylammonium O-isopropyl methylphosphonate salt, 2-hydroxyethylammonium O-isopropyl fluoride, isopropyl 2-aminoethyl methylphosphonate, diisopropyl methylphosphonate, tributylamine, 1,3-diisopropylthiourea, 2-hydroxyethylammonium methylphosphonate salt, other methylphosphonates, benzene, GB (< 25ppb), Methyl ethyl ketone, and trace metals
VX Neutralent	MEA, water, Sodium hydroxide, Sodium 2-diisopropyleaminoethanethiolate, Sodium O-ethyl methylphosphonate, Sodium O-(2-aminoethyl) methylphosphonate, Disodium methylphosphate, Bis-(2-diisopropylaminoethyl) sulfide, Bis-(2-diisopropylaminoethyl) disulfide, 2-diisopropylaminoethyl ethyl sulfide, 1,3-dicyclohexylurea, ethanol, cyclohexylamine, 2-diisopropylaminoethanol, ethylene glycol, N-(2-hydroxyethyl) methylphosphoramidate, VX (< 1ppm), and trace metals

4. Selection of a Transportation and Disposal Strategy for RCWM

The RRS and MMD-1 systems are scheduled to begin full scale testing with chemical agents in 1998. These tests will demonstrate the operational characteristics of these prototype systems and the agent destruction levels achievable with aged chemical agents contained in recovered CAIS and non-explosively configured munitions. The test results will serve several important purposes, including the selection of an overall strategy for transportation and disposal of RCWM as part of the ongoing Non-Stockpile Program Programmatic Environmental Impact Statement (PEIS) process.

In addition to the current strategy of onsite treatment of RCWM and offsite disposal of process wastes, a second strategy, offsite treatment of the CWM, is also under consideration in the PEIS process. A third alternative, onsite treatment of CWM and resulting waste streams, may be evaluated in the future if alternative post-treatment technologies are available in transportable form. Over the course of the next year, the PMNSCM will conduct a detailed evaluation of these strategies, taking into account factors such as environmental impact, socioeconomic impact and risk to the public. Through public meetings and other communication forums, it has become clear that the American public, environmental advocates and other stakeholders are willing to accept these strategies, with some reservations about offsite incineration of process wastes. Of particular concern to some members of the American public is the desire to ensure that commercial incineration can be conducted without adverse effects to the public. Additional data to support such assessments can be gathered through trial burns at U.S. facilities, or, alternatively, may be available from members of the international community with experience in the incineration of equivalent waste streams. In the spirit of international cooperation, the PMNSCM welcomes technical advice on these matters from members of the world chemical demilitarization community. The PMNSCM also welcomes the opportunity to assist the international community with chemical demilitarization programs through technical assistance and information exchange. Technical advice and inquiries should be addressed to:

COL Edmund W. Libby
Project Manager for Non-Stockpile Chemical Materiel
ATTN: SFAE-CD-N, Bldg E4405
Aberdeen Proving Ground, Maryland 21010-5401
Tel. (410) 671-1083 Fax (410) 612-8737
E-Mail: elibby@cdra.apgea.army.mil

5. References

1. Olajos, E. J. and Rao, V., (1997) *Department of Transportation Dermal Toxicity Evaluation of Neutralized CAIS*, Non-Stockpile Chemical Materiel Program Technical Report.
2. Edgewood Research, Development and Engineering Center (ERDEC) (1996) *Evaluation of the Chemistry which Supports the Munitions Management Device-1 (MMD-1) For Chemical Agent HD Neutralization Process, for Program Manager, Non-Stockpile Project*, ERDEC Draft Technical Report.
3. Edgewood Research, Development and Engineering Center (ERDEC) (1996) *Evaluation of the Chemistry which Supports the Munitions Management Device-1 (MMD-1) For Chemical Agent GB Neutralization Process in Both MEA/H2O and KOH/CH3OH/H2O, for Program Manager, Non-Stockpile Project*, ERDEC Draft Technical Report.
4. Olajos, E. J., Salem, H., and Gieseking, J. (1997) *Department of Transportation Dermal Toxicity Evaluation of Solutions Resulting from the Chemical Neutralization of HD, GB, and VX via Monoethanolamine*, ERDEC Technical Report Number ERDEC-SP-052.

HYDROLYSIS AND OXIDATION PROCESS EFFLUENTS OF SOME CHEMICAL WARFARE AGENTS AND POSSIBLE SECONDARY TREATMENTS

RICHARD SOILLEUX
DERA Porton Down

1. Introduction

Traditional chemical decontamination and disposal treatments for chemical warfare (CW) agents rely largely on base hydrolysis or oxidation reactions in aqueous solution. Although often used as methods for decontamination or for disposing of relatively small quantities of agent (often as a partial treatment together with open pit burning), these types of process have also been used in the US, UK and Canada, for example, to dispose of relatively large amounts of CW agent. The use of these chemical neutralisation treatments for large scale disposal was superseded in recent times by the use of incineration. In the 1950's, large quantities of mustard were destroyed by incineration (which is of course an oxidation reaction) in the UK. More recently, the Canadian stockpile was also disposed of by incineration as were the Iraqi stocks of mustard by UNSCOM. The US baseline technology is also, of course, based on incineration and this, together with the existing incinerators in Germany and the UK, uses modern pollution abatement systems and is efficient and effective. Nevertheless, there is considerable public opposition to incineration and alternative technologies are actively being sought in several countries.

Currently, neutralisation methods are being considered as alternatives to incineration, mainly in the US and FSU, for the bulk disposal of CW agents. Although the basic processes of base hydrolysis and oxidation have been used traditionally, some of the alternative technologies currently being proposed to use them are novel and sophisticated. They include silver II, supercritical water oxidation, hot water hydrolysis/bioremediation and molten salt oxidation. This paper concentrates on aqueous based hydrolysis and oxidation reactions, their effluents and effluent treatments.

I do not propose to talk about the treatment of WWI volatile agents such as CG or CK which, although they are extremely hazardous when encountered in the field, do not present a large scale disposal problem. Instead, I will concentrate on CW agents that are in current stockpiles or regularly recovered in site clearance operations

F.W. Holm (ed.), *Effluents from Alternative Demilitarization Technologies*, 27–33.

2. Traditional Methods of Treatment of CW Agents by Hydrolysis

Table 1 gives the hydrolysis products of selected CW agents. As expected from hydrolysis reactions, these products include complex organic molecules that need further treatment before disposal. Hydrolysis should, therefore, be regarded as a pre-treatment rather than a complete disposal method. However, hydrolysis has the significant advantage that in most cases CW agents so treated are de-toxified, and, thus, they are much safer to handle. This is true of mustard and the most nerve agents but some of the hydrolysis products of VX and the arsenical CW agents are themselves toxic and still require to be handled with care. It should be noted that lewisite (III) is not susceptible to hydrolysis and can be steam distilled without decomposition. It is, however, much less toxic than the other forms of lewisite and is present, together with lewisite (II), as an impurity in the industrial grade product.

A further feature of the hydrolysis process is that the products of reaction are dissolved in large volumes of water so the quantities of effluent from this partial treatment are much larger than the original amounts of CW agent.

TABLE 1. Hydrolysis products of some CW agents

CW AGENT	pH	HYDROLYSIS PRODUCTS
MUSTARD	7 heat	THIODIGLYCOL, HCl, SMALL QUANTITIES OF COMPLEX THIOLS
MUSTARD	10	THIODIGLYCOL, COMPLEX THIOLS, HCl
SARIN	7	METHYLPHOSPHONIC ACID, HF
SARIN	10	DISODIUM METHYL PHOSPHONATE, ISOPROPYL ALCOHOL, NaF
SOMAN	7	PINACOLYL METHYLPHOSPHONATE, HF
VX	3	ETHYL METHYLPHOSPHONIC ACID, AMINOETHYLTHIOL
VX	7	(S-[2-DIISOPROPYLAMINOETHYL] METHYLPHOSPHONOTHIOIC ACID) i.e., toxic EA 2192
VX	10	ETHYLMETHYLPHOSPHONIC ACID, DIISOPROPYLAMINOETHANETHIOL
DIPHENYL-CYANOARSINE	10 heat	DIPHENYL ARSENIC ACID, HCN
LEWISITE (I)	7	2-CHLOROVINYLARSINE OXIDE, HCl
LEWISITE (II)	7	BIS-(2,2-DICHLORODIVINYLARSINE)OXIDE, HCl
LEWISITE (III)		NOT HYDROLYSED

In some cases, notably that of sarin, the rate of hydrolysis is highly pH dependent and is also accelerated by copper chelates, amines and acetate as well as oxidising agents such as peroxide. (The effects of oxidising reactions will be discussed later.) For example, in non-buffered solution the half life of GB is 1min at pH 10 and 15 min at pH 1 whereas at pH 7 it is 50 hrs. A similar effect is seen with soman. Also, mustard hydrolyses very slowly at low pH, more quickly at pH 7 (99% in 110 min at 20°C) and the rate is about 20% faster than this at high pH. Heat accelerates the hydrolysis of mustard to a marked degree. The rate determining process for the hydrolysis of H is the rate of solution and if this is accelerated by vigorous stirring then the rate of hydrolysis of H in hot water alone is acceptable (i.e., a 4% solution at 90°C is hydrolysed completely in about 30 min. The addition of sodium or calcium hydroxide to the hydrolysate prevents the back reaction and neutralises the HCl formed as the other product of reaction.

2.1. TREATMENT OF EFFLUENT STREAMS FROM HYDROLYSIS

Traditionally, CW agents treated by hydrolysis were regarded as de-toxified and disposed of into lagoons, where they were allowed to concentrate by evaporation and soak into the ground, were dumped at sea, or smaller quantities were simply poured down drains. This is no longer acceptable.

During the difficult conditions faced by UNSCOM in Iraq, the decision was taken to destroy the nerve agent (GB/GF mixture) by hydrolysis. This was done using a 15% solution of sodium hydroxide and resulted in five times the original nerve agent volume of hydrolysate. The hydrolysate was then evaporated to dryness in the desert sun and the mixture of organic and inorganic salts sealed into concrete.

The UNSCOM expedient method is preferable to the traditional methods of disposal following hydrolysis, but further treatment is ideally required to remove the remaining organic material (completely oxidise or mineralise) from the hydrolysates before removal of the remaining inorganic toxic salts for disposal as ordinary toxic waste. As in the UNSCOM method, this can be done by evaporation, or alternatively, by some process that removes the salts from solution to the point where the final aqueous waste stream can be released into the environment, possibly via a sewage treatment plant.

3. Traditional Methods of Treatment of CW Agents by Oxidation

The oxidation of CW agents that takes place with traditional decontamination methods based on hypochlorite results in rapid detoxification which is why it is still used as the main method by the military for the decontamination of equipment in the field. It is not a very attractive method for demilitarisation purposes, however, since (like base hydrolysis) it does not result in complete demineralisation, and it also adds significantly to the environmental burden of the waste stream.

With many agents, reactions with dry bleach (calcium hypochlorite/ calcium oxide) can be violent and mustard will catch fire if treated in this way. Bleach is, therefore, used as a decontaminant in the form of an aqueous slurry. However, problems occur due to settling out and the difficulties of getting the solid bleach to mix efficiently with the liquid agent, especially for those that have very low solubility's in water. This, however, does have the advantage, over sodium hypochlorite, that the calcium salts that are the products of the oxidation reaction are also relatively insoluble and can be readily removed from the effluent stream. Bleach also has the disadvantage that it is extremely corrosive to equipment and loses chlorine, and hence effectiveness on storage. It is, however, widely available and cheap.

Oxidation reactions with bleach have been included for completeness, but, because of the environmental and other problems associated with its large scale use it will not be considered further here.

Foremost among other possible oxidising agents for CW agents that are more environmentally benign than bleach is hydrogen peroxide. This is a powerful oxidising agent which does not add to the environmental burden and introduces only water into the waste stream. It has not been used widely for demilitarisation of CW agents although it has been used together with sodium hydroxide for the disposal of L and VX. A disadvantage is that reactions with concentrated solutions can be violent. Ozone is another powerful oxidising agent that does not add to the environmental burden; there is, however, little experience of it being used for demilitarisation. Another disadvantage of ozone is that it cannot be stored ; it must be generated *in situ*.

TABLE 2. Products of traditional oxidative decontamination treatments of some CW agents

CW AGENT	OXIDISING AGENT	PRODUCTS OF OXIDATION
MUSTARD	HYPOCHLORITE	BIS(2-CHLOROETHYL)SULPHOXIDE, SULPHONE, VINYL COMPOUNDS
MUSTARD	HYPOCHLORITE/ BURNING	SO_2, CO_2, HCl, H_2O, 1,4 DITHIANE, 1,4 DIOXANE
SARIN	HYPOCHLORITE	ISOPROPYLMETHYL PHOSPHONIC ACID
SOMAN	HYPOCHLORITE	PINACOLYLMETHYL PHOSPHONIC ACID
VX	HYPOCHLORITE	ETHYL METHYLPHOSPHONIC ACID, O-ETHYL S-2-(DIISOPROPYL AMINO) ETHYLMETHYLPHOSPHONOSULPHOXIDE
DIPHENYL-CYANOARSINE	HYPOCHLORITE	DIPHENYLARSONIC ACID, HCN
LEWISITE (I)	HYPOCHLORITE	ARSENATE, ACETYLENE
LEWISITE (II)	HYPOCHLORITE	ARSENATE, ACETYLENE
LEWISITE (III)	HYPOCHLORITE	ARSENATE, ACETYLENE

4. Mineralisation of waste streams from hydrolysis and/or oxidation reactions

Complete oxidation occurs when efficient combustion takes place in a modern incinerator. This process is often also referred to as mineralisation because it results in the complete oxidation of all organic moieties to inorganic molecules. Table 3 shows the products of complete oxidation of the CW agents under review. The products consist largely of carbon dioxide and water with acid gases NOx and SOx together with involatile or low volatility acids, such as, phosphoric and arsenic acids.

TABLE 3. Products of complete oxidation of some CW agents

CW AGENT	PRODUCTS OF COMPLETE OXIDATION
MUSTARD	SO_2, HCl, CO_2, H_2O
SARIN	H_2PO_4, HF, CO_2, H_2O
SOMAN	H_2PO_4, HF, CO_2, H_2O
VX	H_2PO_4, SO_2, CO_2, NO_3, H_2O
DIPHENYL-CYANOARSINE	$H_2As_2O_6$, CO_2, H_2O, NO_2
LEWISITE (I), (ii) and (III)	$H_2As_2O_6$, CO_2, H_2O, HCl

Complete oxidation is the ultimate goal of all the CW agent demilitarisation reactions discussed above. Once this is achieved, the final stage is to concentrate the resulting brines for efficient conventional toxic waste disposal.

Several methods have been proposed to mineralise the effluent streams from hydrolysis and/or oxidation processes. The most obvious of these methods is incineration, which is undoubtedly effective, and, with a modern pollution abatement system, pollutants are removed efficiently from the effluent streams. The advantage of incinerating effluents from chemical detoxification processes is that the incinerator does not need to be built or operated at the high safety standards necessary for the direct incineration of agent. Also, small amounts arising from non-stockpile EOD operations can in principle be detoxified on-site and transported safely to a central location for disposal. The main disadvantages of this approach are the additional fuel costs of incinerating large quantities of water and the public objections to incineration in any form.

Some other possible mineralisation treatments for the effluents of hydrolysis or oxidation reactions are silver (II), molten salt, and molten metal. Molten salt and, especially, molten metal are both high temperature treatments and may suffer, as does incineration, from adverse public reaction. Silver (II), on the other hand, has inherent safety features and is a low temperature ambient pressure process. Disadvantages for all three processes for the treatment of dilute effluent streams is that they are all complete processes in their own right and would be more efficiently used to destroy CW agents directly. Silver (II), however, can deal efficiently with reasonably dilute feeds (e.g., 20%) and, indeed, requires water to be added to the analyte to replace the water transported across the membrane by the current. Therefore, silver (II) is a future contender for the treatment of hydrolysate. However, it has not yet been proven in this role, and, along with the other possible future methods, it will not be discussed further.

To date, two alternative mineralisation processes have been selected for the US CW stockpile disposal program. These are bioremediation, following hot water hydrolysis of the HD stockpile at Edgewood Arsenal and super critical water oxidation (SCWO) for hydrolysed VX from the stockpile at Newport Chemical Depot. The advantages of SCWO and bioremediation for the mineralisation of hydrolysates are that they both require a dilute aqueous phase feed and are, therefore, complementary to base hydrolysis. They are probably the best available techniques, at present, for the treatment of hydrolysates and will now be discussed in a little more detail.

4.1. BIOREMEDIATION

Bioremediation removes the organic material and some of the inorganic salts from solution and incorporates them into biomass. The waste water emerging from a properly configured bioreactor can then be released directly into the environment.

Depending on the origin of the hydrolysate, excess biomass can either be returned directly to the environment, dewatered and land filled, or it may need further treatment. For example, although biomass derived from mustard should be perfectly safe. it could produce bad odors due to the release of sulphur compounds, and it may have a higher than normal sodium chloride content. Biomass derived from nerve agents would add phosphorous, which is a plant nutrient, to the environment and may affect the balance of fertility in the soil and/or water courses.

The above problems are relatively minor and can be dealt with by careful consideration of the eventual fate of the effluent, dilution and mixing of waste streams, etc.. A more difficult problem arises, however, with biomass derived from the hydrolysate from arsenical CW agents. The microorganisms may either retain the arsenic and require further treatment or release inorganic arsenic compounds in water soluble form.

There is also the possibility that volatile arsenic compounds could be produced, which could be a problem, since, the gaseous emissions from a bioreactor are normally considered as environmentally benign. This, however, is unlikely in the aerobic systems currently under study, although it could be a problem with anaerobic systems. The production of volatile arsenic compounds could conceivably be turned to advantage as a convenient method for removing this toxic element from solution.

This aspect, of arsenic entrapment and removal, needs further investigation before bioremediation can be recommended as a mineralisation treatment for the hydrolysate from arsenical CW agents.

This method has great potential since it uses little power, is similar to traditional sewage treatment technology, and is seen by the public as a low - tech natural process that is environmentally benign.

4.2. SCWO

SCWO as envisaged for the mineralisation of VX hydrolysate at Newport will take the diluted and oxygenated waste stream and heat it under pressure to achieve supercritical conditions for water. Operating conditions are typically 600° C and 250 bar when organic compounds are rapidly and completely oxidised with the formation of carbon dioxide and mineral salts. Gaseous emissions are normally low and SOx and NOx and PAH's are not a problem. Volatile arsenic compounds are unlikely to be produced in the highly oxidising conditions and this element should remain in the effluent solution as mixed salts of arsenic acid. The dilute mineralised effluent must be concentrated, probably by evapouration, for efficient disposal of the salts.

This technology is efficient and effective but there are technical problems in that there is a high consumption energy and corrosion can be severe in the reactor vessel which must be made of high strength corrosion resistant materials such as hastelloy.

5. Conclusions

Hydrolysis and oxidation reactions have been reviewed for selected CW agents. In general, hydrolysis results in detoxification of the agent, but retains the bulk of the organic material in dilute aqueous solution as a large molecule requiring further treatment before release into the environment.

Oxidation using the traditional decontamination method with bleach slurry is effective for detoxification; however, it adds considerably to the environmental burden and is not recommended for demilitarisation.

Oxidation using hydrogen peroxide is environmentally more benign than bleach and can be used in conjunction with base hydrolysis to improve the extent of detoxification with intransigent CW agents such as L(III). However, under normal conditions, mineralisation does not occur and further treatment, such as that recommended for hydrolysate waste streams, is required.

Complete oxidation (mineralisation) is the ultimate goal of all CW agent demilitarisation reactions. Suitable treatments for the effluents from hydrolysis and/or oxidation reactions of CW agents have been discussed and bioremediation and SCWO identified as the most promising at present.

The aspect of arsenic entrapment and removal needs further investigation before bioremediation can be recommended as a mineralisation treatment for the hydrolysate from arsenical CW agents.

Once mineralisation has been achieved the final stage is concentration of the resulting brines for efficient conventional toxic waste disposal.

6. References

1. Review of Suspected Warfare-Related Environmental Damage in the Federal Republic of Germany (2nd Ed.) Part 3: Chemical Agents Dictionary, P Kopecz in collaboration with J Thieme, Pub. Umweltbundesampt, Postfach 33 00 22 D - Berlin, Germany, (1996).
2. Chemical Reactions for Neutralising Chemical Warfare Agents, Yu-Chu Yang, Chemistry & Industry, 334 - 337, 1 May (1995).
3. Decontamination of Chemical Warfare Agents, Yu-Chu Yang, J A Baker, and J R Ward, Amer. Chem. Soc., Chem. Rev. 1729 - 1743 (1992)
4. Alternative Technologies for the Destruction of Chemical Agents and Munitions, Committee on Alternative Chemical Demilitarisation Technologies, Board on Army Science and Technology Commission on Engineering and Technical Systems, national Research Council, Washington D.C. 1993.

PILOT-SCALE BASE HYDROLYSIS PROCESSING OF HMX-BASED PLASTIC-BONDED EXPLOSIVES

RAYMOND FLESNER, P.C. DELL'ORCO, T. SPONTARELLI, R.L. BISHOP, C.B. SKIDMORE, K. UHER, AND J.F. KRAMER
HE Science and Technology, Los Alamos National Laboratory
P.O. Box 1663, m/s C920, Los Alamos, NM 87545

1. Abstract

Los Alamos National Laboratory has demonstrated that many energetic materials can be rendered non-energetic via reaction with sodium hydroxide or ammonia. This process is known as base hydrolysis. A pilot scale reactor has been developed to process up to 20 kg of plastic bonded explosive in a single batch operation. In this report, we discuss the design and operation of the pilot scale reactor for the processing of PBX 9404, a standard Department of Energy plastic bonded explosive containing HMX and nitrocellulose. Products from base hydrolysis, although non-energetic, still require additional processing before release to the environment. Decomposition products, destruction efficiencies, and rates of reaction for base hydrolysis will be presented. Hydrothermal processing, previously known as supercritical water oxidation, has been proposed for converting organic products from hydrolysis to carbon dioxide, nitrogen, and nitrous oxide. Base hydrolysis in combination with hydrothermal processing may yield a viable alternative to open burning/open detonation for destruction of many energetic materials.

2. Introduction

2.1. BENCH AND PILOT-SCALE BASE HYDROLYSIS OF HMX BASED EXPLOSIVES

The safe disposal of excess high explosives has become increasingly important in recent years due to the disassembly of large quantities of Cold War munitions. Thus far, the primary disposition of waste high explosives has been open burning/open detonation (OB/OD). Due to noise concerns and the dispersion of undesirable burning products, OB/OD has recently fallen out of favor with state and federal regulatory agencies. This has prompted the development of safe and environmentally benign disposal alternatives. One high explosive disposal alternative being developed in parallel by industry and Los Alamos National Laboratory is base hydrolysis.

F.W. Holm (ed.), Effluents from Alternative Demilitarization Technologies, 35–45.

Base hydrolysis is a process whereby explosives are reacted with concentrated base solutions (1-8 M) at elevated temperatures (80°C to 150°C) to produce non-energetic, water soluble reaction products. At temperatures above the normal solution boiling point, the process is carried out in pressurized reactors. Bases used include sodium hydroxide, potassium hydroxide, ammonium hydroxide, and sodium carbonate; although water is the most commonly used solvent, other solvents (i.e., ethanol, DMSO) may be used neat or in combination with water to reduce mass transfer resistances to reaction. Several different reaction paths occur in the hydrolysis process, including hydrolysis, oxidation/reduction, and substitution/elimination. These reaction paths produce a product stream which largely consists of simple carboxylic acids, amines, sodium salts, ammonia, and gaseous nitrogen/nitrous oxide. In general, hydrolysis reactions consist of parallel mass transfer and reaction kinetic steps; the temperature, pressure, and solvent combination determine whether mass transfer or reaction kinetics is the rate limiting step.

Many types of explosives have been successfully treated using base hydrolysis.[1] Both nitramines and nitroaromatics are converted to water soluble, non-energetic products. When high volume loadings of base-resistant binders are present in a formulation, solvents generally have to be added to swell the binder, thus allowing base access to the high explosive. In addition to bulk explosives, base hydrolysis has also been shown to be effective in regenerating explosive-laden activated carbon,[2] and in treating HE contaminated rags and soils.

Base hydrolysis has been successfully transitioned to large scale operations. Currently, two large-scale units are scheduled for implementation later this year at Eglin Air Force Base and the Pantex Plant in Amarillo, Texas An ammonia-based hydrolysis process is operational at Thiokol to treat CYH rocket propellant .[3] In addition, Chemical Systems Division, a propellant manufacturer in northern California, is currently using base hydrolysis to destroy some of their waste energetic materials.[4] The objective of this paper is to summarize research efforts in the base hydrolysis of HMX formulations at Los Alamos National Laboratory (LANL). Research at LANL has been primarily focused on PBX's from Department of Energy dismantlement programs. Specifically, we have focused on PBX 9404, a 94% HMX composition. Both bench and pilot-scale studies have been conducted, with the purpose of developing a prototype reactor for use at the DOE's dismantlement plant at Amarillo, Texas. Atmospheric and pressurized hydrolysis processes have been evaluated in the laboratory. Results from the testing of a pilot-scale atmospheric hydrolysis reactor for PBX 9404 will be presented here. Ongoing work includes further investigation of mass transfer resistances as a function of temperature, pressure, and base composition, and the development of a pilot-scale pressurized hydrolysis reactor for use in DOE dismantlement activities.

3. Experimental Methods

A pilot-scale batch reactor was constructed for the purpose of hydrolyzing large explosive pieces into water soluble, non-explosive products. PBX 9404 (94% HMX, 3% nitrocellulose binder, 3% chloroethylphosphate plasticizer, and 0.1% diphenylamine (nitrocellulose stabilizer)) was used in this study. Two forms of explosive were studied: molding powder and consolidated charges. Molding powder consisted of small particles or pieces of plastic-covered explosive up to 0.5-1 cm in diameter. Consolidated pieces were manufactured by pressing molding powder in high pressure isostatic presses at elevated temperatures.

A stainless steel 55-gallon container was used as the basis of the reactor system. Two 7.5 KW Tempco electric circulation heaters (model # CR3N-4008-4G 7.5) were used to heat the fluid (NaOH, 0.8-1M) up to 90°C. Hydrolysis liquor was circulated through the heaters using a Wilden M-1 stainless steel/Teflon air operated diaphragm pump (model # M1/ST/TF/TF/ST). Hydrolysis liquor was pumped from the bottom of the reactor, into the diaphragm pump, through the heaters, and then back into the sides of the reactor. A schematic of the initial reactor configuration is shown in Figure 1. The flow pattern and design of the reactor were later modified to alleviate solid accumulation problems, as discussed later.

An experiment was initiated by filling the reactor with a base solution. Either molding powder or a consolidated piece was placed in a stainless steel basket inside of the reactor. For the molding powder experiments, only one basket was used. For the first consolidated piece experiment, a cage was used to initially place the explosive inside the reactor. This cage rested on the bottom of the steel basket inside the reactor. In an attempt to increase the mass transfer in the reactor and therefore the overall reaction rate, the steel basket was not used in the second consolidated piece experiment. The steel cage merely rested on the bottom of the reactor immediately above an air sparge.

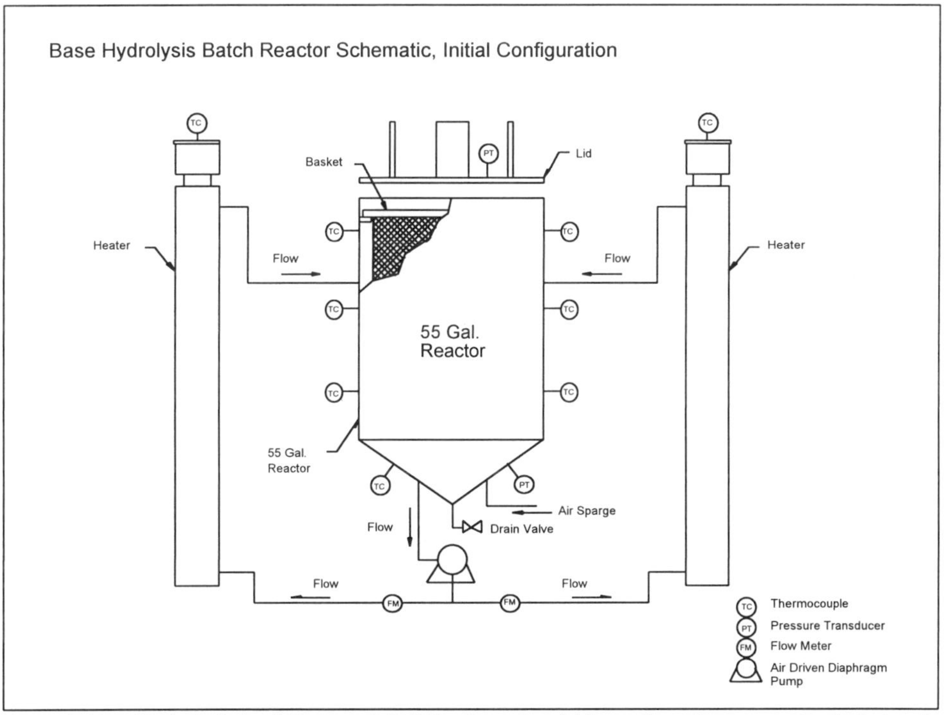

Figure 1. Base Hydrolysis Batch Reactor, Initial Configuration. This configuration was used only during the first 30 pound molding powder experiment.

Mixing in the reactor was accomplished by pump circulation and an air sparge. One inch stainless steel tubing and Swagelok fittings were used as plumbing between the reactor, pump, and circulation heaters. Paddlewheel flowmeters (Omega, model #FP-5300) measured flow through either side of the reactor. Type K thermocouples were used to monitor temperature at various points in the reactor and in the heater's thermowells. Viatran pressure transducers (model 244) measured head pressure of the liquid inside the reactor.

The system was monitored remotely during experiments using a Stanford SR630 thermocouple monitor, an Apple Macintosh computer with a National Instruments GPIB board, and National Instruments LabVIEW software. Thermocouple, flow, and pressure measurements were all monitored at the computer, and this data was all logged to the hard drive as an ASCII file. In addition, the LabVIEW program controlled the temperature of the heaters via a PID control scheme that fed a 4-20ma current to 3 phase, 208V Omega solid state relays (SCR73Z-260 60). The control panel from the LabVIEW program is shown if Figure 2.

Liquid samples were analyzed by ion chromatography (IC) for anions and cations, total organic carbon (TOC) and total inorganic carbon (TIC), and by a pH meter. The IC anion analysis determined acetate, formate, nitrite, nitrate, oxalate, and glycolate. Cations measured were ammonia, sodium, and methylamine.

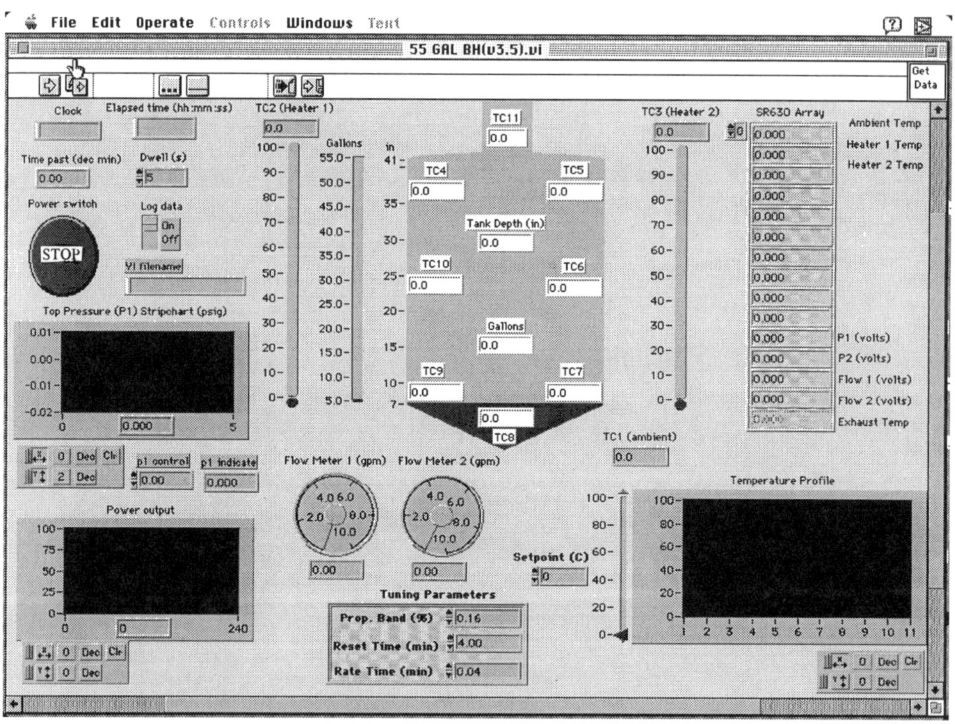

Figure 2. LabVIEW Control Panel for Computer Control of Pilot-Scale Base Hydrolysis Unit. LabVIEW was run on an Apple Macintosh IIfx computer.

4. Results and Discussion

4.1 PILOT-SCALE RESULTS

4.1.1. *PBX 9404 molding powder*

Two experiments were performed using 30 lb of molding powder. In the first experiment, significant solids accumulation occurred. This accumulation severely choked fluid flow through the heaters. In the original plumbing scheme (Figure 1), the powder was pumped from the bottom of the reactor, into the pump, and then through the external circulation heaters. The nitrocellulose in the powder reacts relatively fast (within minutes at 80°C),[5] and most of the HMX accumulated in the pumps and the plumbing near the bottom of the reactor. Within the first 40 minutes of operation, the reactor plugged. The bulk fluid temperature at the time of plugging was 68°C. Although serious plugging occurred throughout the bottom of the reactor system, no serious exotherms were detected in any part of the reactor.

Because of the serious nature of the solids accumulation, the reactor was reconfigured so that fluid was pumped from the side of the reactor, into the diaphragm pump and to the heaters before being returned to the bottom of the reactor. The new pumping scheme is shown in Figure 3. The advantage of this pumping scheme is that HMX

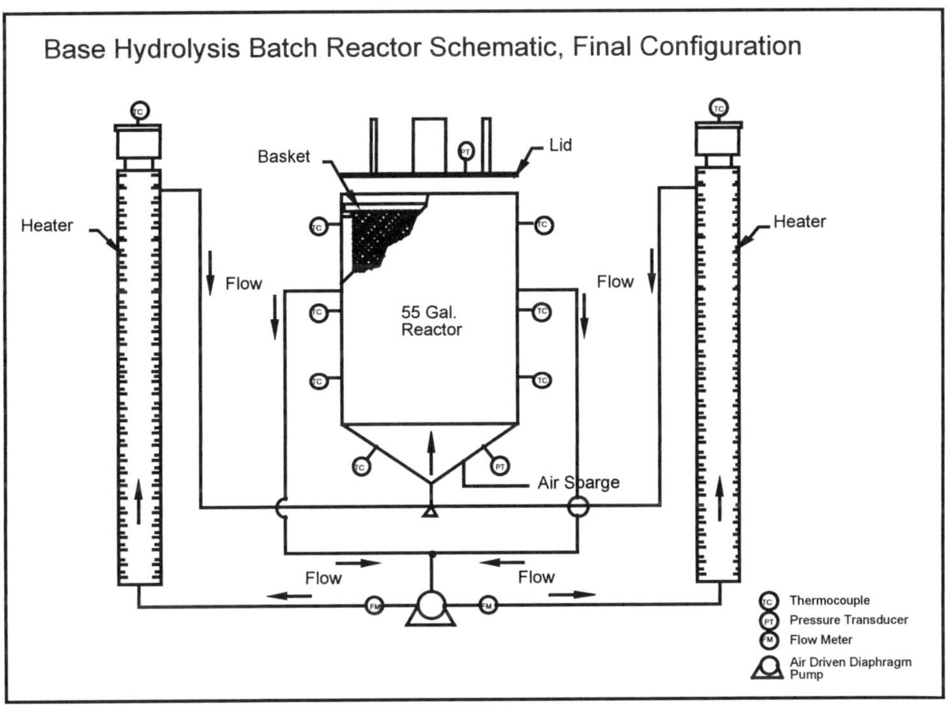

Figure 3. Base Hydrolysis Reactor Schematic, Final Configuration. This configuration was used after the reactor plugged significantly in the first molding powder experiment.

powder that has not yet reacted in the tank will not settle and plug the reaction pumping system and heaters. The disadvantage is that since the pump inlet is taking fluid from the sides of the reactor, a minimum amount of liquid must be maintained in the reactor for the system to work properly. This minimum amount is approximately 40 gallons.

Another 30 lbs molding powder experiment was conducted with the configuration displayed in Figure 3. In this second molding powder experiment, the fluid was heated from 30°C to 85°C in approximately 1.6 hours. Similar to the first experiment, by the time the solution reached 85°C, most of the nitrocellulose had reacted releasing the fine HMX powder. At this point, possible intermittent plugging was observed in one of the flow lines, which continued for about 70 minutes. Instead of plugging, the paddle of one of the flowmeters may have caught itself during this time on a plastic insertion fitting. The reaction time-temperature history is shown in Figure 4. In this figure, the heater temperature, the bulk fluid temperature, and the set point temperature are shown. Approximately 2 hours after the experiment began, one of two flow sensors indicated that intermittent plugging or flowmeter measurement problems were taking place. When this was observed, the heaters were completely turned off. This is indicated by the abrupt change in temperature set point at approximately 3 hours time, (dashed line, Figure 4). After the heating power was turned off, a slight exotherm of approximately 4°C was also observed and is indicated on the graph. Most likely, some nitrocellulose continued to hydrolyze and caused the temperature increase.

4.1.2. *PBX 9404 consolidated pieces*

This reactor was originally designed to remediate large pressed pieces of PBX 9404, not molding powder. Much of the DOE's remediation mission is to get rid of large pressed pieces. These pieces require a longer reaction time since the system is mass transfer limited, which leads to less powder present at any one time, reducing the chance of plugging. Furthermore, the slower reaction rate will cause the exotherm created by the degradation of nitrocellulose to be much less. The molding powder experiments were judged to be a worst case reaction scenario for solids loading in the reactor. They were carried out to both demonstrate the robust capabilities of the system and to give the system a thorough safety test.

Consolidated pieces were hydrolyzed in two different experiments. Each experiment involved a single pressed PBX 9404 piece made from molding powder. In the first experiment, a charge weighing less than 35 lb was hydrolyzed in 45 gal of 1 M NaOH. The fluid was initially heated to near 90°C and then held above 80°C for 9 hours. The reactor was allowed to cool overnight. The next morning, the basket was taken out of the reactor, and a significant quantity of the pressed piece remained. The piece was not weighed at this point, but was returned back to the reactor and heated again that day. The reactor was held above 80°C for an additional 8.7 hours. The reactor was then allowed to cool overnight. The time-temperature history of the reaction is shown in Figure 5. The following morning, a small piece was removed. The solution was then drained and filtered, and a small amount of additional HMX powder was recovered. The total amount left was 6% of the starting material. The rest of the explosive had been converted to base hydrolysis products (i.e., water soluble aqueous compounds, and N_2, N_2O, and NH_3).

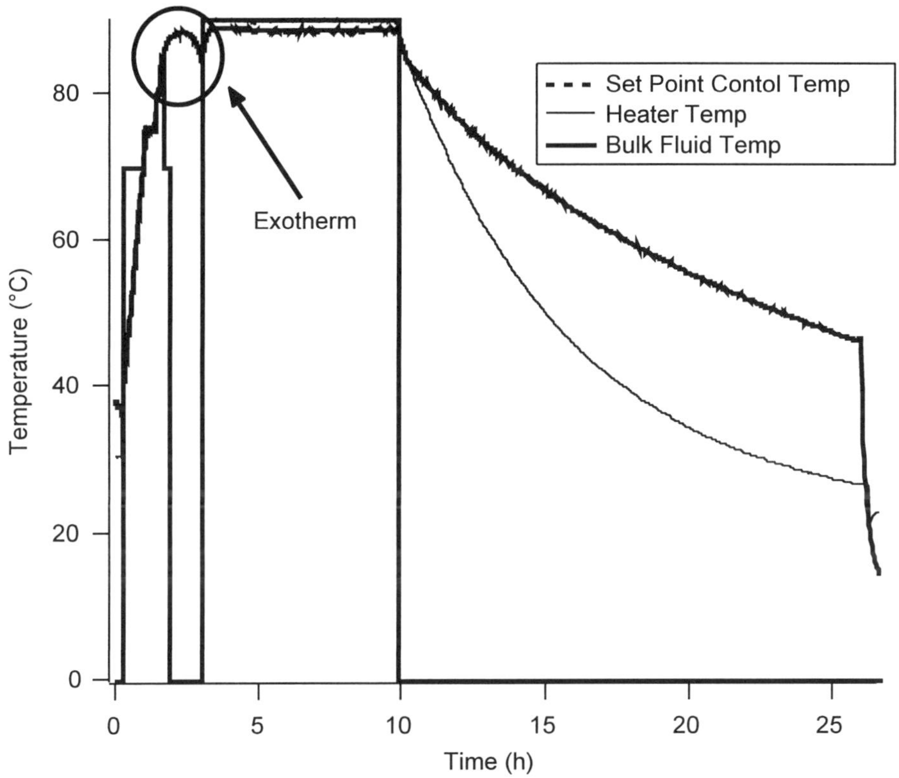

Figure 4. Second PBX 9404 Molding Powder Experiment. Thirty pounds of PBX 9404 molding powder were hydrolyzed in 45 gal of 1.0 M NaOH.

In the second experiment, we attempted to increase the mixing and therefore the overall mass transfer of the solid explosive into the surrounding base solution. We achieved this by removing the basket that held the explosive as previously described in the Experimental Methods section. It was hoped that the additional mixing created by having the air sparge near the explosive would cause the explosive to hydrolyze faster. This was not necessarily the case. In the second experiment, we achieved approximately 87% destruction. This was less than the first experiment, where we achieved 94% destruction, however the reaction was not held at temperature as long. These results are summarized in Table 1.

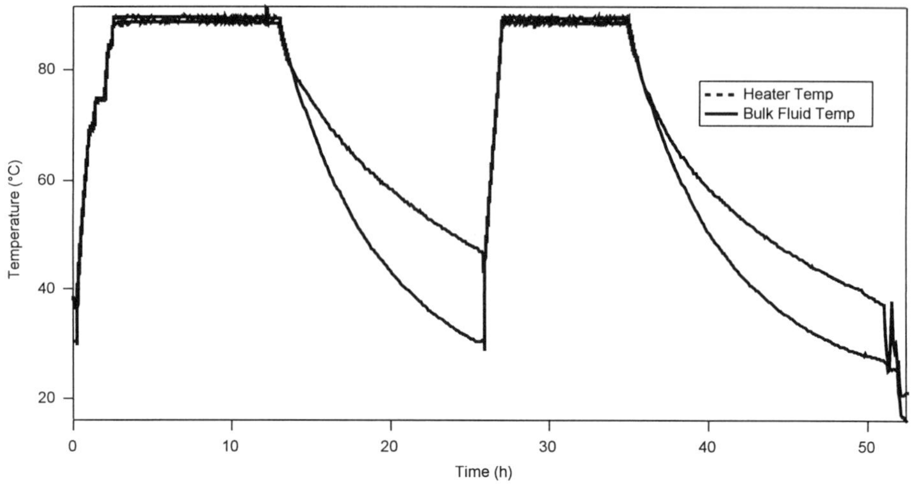

Figure 5. First PBX 9404 Consolidated Piece Experiment. One PBX 9404 pressed piece (weighing less than 35 lb) was hydrolyzed in 45 gal of 1.0 M NaOH.

Table 1. Destruction percentage of consolidated PBX 9404 High Explosive		
Experiment Number	Time above 80°C (h)	High Explosive Destruction %
Consolidated Exp #1	21	94
Consolidated Exp #2	18	87

Both experiments proceeded at approximately the same rate. If the rate is assumed to be approximately linear for the whole reaction, the amount of explosive hydrolyzed per unit time in the second experiment is slightly more (4.9% reacted per hour vs. 4.6% for the first experiment). This assumption is not completely true; past experiments have shown that the reaction rate is not linear with time.[5] The reaction slows somewhat as it proceeds, both due to the depletion of hydroxide ion as the reaction progresses and to the increased concentration of soluble salts in the liquid that tend to "salt out" organic species, thus reducing their solubility. Approximate reaction rates of the two consolidated pieces are reported in Figure 6. Also shown in Figure 6 is the overall reaction rate of the second molding powder experiment and a comparison to experiments performed in a lab scale reactor at 90 and 150°C. In one of the lab-scale experiments (indicated by 150-P on the x-axis), a single consolidated piece (2" diameter by 3" high cylinder) weighing a total of 284 g was used. In this experiment, the temperature was raised to 150°C for 53 min. At the end of the experiment, 99.8% of the explosive had been destroyed.

Figure 6 shows a qualitative comparison in reaction rates between molding powder and consolidated pieces at both small and large scales. Since this is a mass transfer limited reaction,[6] the available explosive surface area plays a large role in the rate. This is reflected in the faster reaction rate for molding powder compared to the consolidated pieces. The high temperature experiments are included to show the increase in mass transfer (even without stirring), and therefore the overall increase in reaction rate. A direct comparison of rates for consolidated pieces shows that destruction rates increase nearly 2 orders of magnitude when the temperature is increased from 90°C (mixed) to 150°C (with no mixing). We do not expect that reaction rates will scale linearly with respect to explosive size because of the general decrease in ratio of surface area to volume when larger consolidated pieces are used. Preliminary estimates indicate that a large scale charge may take from 2 to 4 times longer than the small scale piece treated at 150°C. Mixing should also significantly increase mass transfer at the higher temperatures as well.

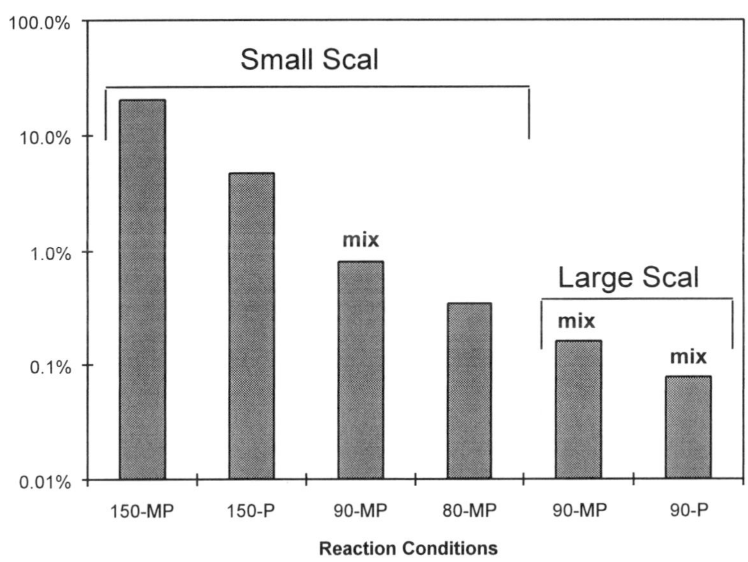

Figure 6. Reaction rate comparison between small and large scale reactions. Temperature (°C) and explosive form are indicated on the x-axis. MP refers to molding powder and P refers to consolidated pieces. Mixing was carried out in some experiments and is indicated.

4.1.3. *Product Analysis*

Major carbon products in solution consist of sodium formate and sodium acetate. These two products account for approximately 40 to 50% of the total carbon from the high explosive. Sodium nitrite is the major nitrogen product in solution accounting for approximately 20% of the nitrogen in the explosive. Gas data were not collected in these experiments. In previous experiments,[6] nitrogen gas, nitrous oxide, and ammonia are released in the reaction, and account for about half of the nitrogen in the explosive. Total carbon and total organic carbon were also measured at the end of each experiment. These results are summarized in Table 2, which presents Ion Chromatography results for both sets of experiments. We are also aware that sodium glycolate is present in our product mixture, but glycolate and acetate elute out at similar retention times in our IC analysis.

The products shown in Table 2 are amenable to treatment by hydrothermal oxidation. Indeed, we have had success mineralizing organic carbon using nitrite and nitrate available in the solution.[1] Hydrothermal oxidation experiments performed on PBX 9404 hydrolysate indicated that organic carbon conversions greater than 99% were achieved at relatively mild operating conditions (374°C, 300 bar). To completely mineralize the organic carbon, additional oxidant such as hydrogen peroxide must be added to the hydrothermal system.[1]

Table 2. Ion Chromatography Results for PBX 9404 pilot scale hydrolysis.

IC products (% of C or N in explosive processed)	Molding powder 2	Consol. piece 1	Consol. piece 2
Acetate	19%	22%	25%
Formate	16%	21%	23%
Nitrite	20%	19%	20%
Nitrate	0.7%	0.5%	0.7%
Organic carbon	99%*	93%	101%
Inorganic carbon	1.1%*	1.4%	1.8%

- Mass balances based on Total Carbon analytical results rather than amount of explosive left at end of experiment.

4.2 PRESSURIZED REACTOR SCALE-UP

Los Alamos is in the process of constructing a new pilot scale reaction vessel that will incorporate a pressurized base hydrolysis semi-batch reactor and a hydrothermal reactor that will process the base hydrolysis product. This system will remediate HMX-based plastic bonded explosives. The major aqueous and gaseous products from this reactor system will be primarily sodium carbonate, nitrous oxide, and nitrogen. The base hydrolysis reactor will be run in semi-batch mode with concentrated sodium hydroxide fed continuously during the reaction. Thirty to fifty pounds of explosive will be able to be processed in the reactor during a run, and residence times should be on the order of 1 to 2 hours per run.

Safety studies are currently in progress to determine the maximum safe operating temperature for the base hydrolysis of consolidated pieces. Small scale studies have been operated at temperatures up to 150°C with no severe or uncontrollable reactions. This may not necessarily be the case for large consolidated pieces of explosives. Runaway reactions may occur if heat buildup is not adequately taken into account and cannot be removed fast enough. The results of the safety studies should identify safe operating temperatures for pressurized base hydrolysis batch reactions at large scale, and these results will be reported when the experiments are complete.

5. Conclusions

We have successfully hydrolyzed HMX plastic bonded pressed explosives at a pilot scales of less than 35 lb of explosive. Reaction rates are slower than similar bench scale experiments due to decreases in mixing efficiency at larger scales. Times up to 20 hours are required to destroy 90% of the explosive. Major aqueous carbon products include sodium acetate, formate, and glycolate, all of which are amenable to hydrothermal oxidation. Lab scale experiments at higher temperatures (e.g., up to 150°C) indicate that destruction rates may be an order of magnitude faster than reactions carried out at 90°C. More work is needed to quantify scale-up data for these higher temperatures.

6. Acknowledgments

We would like to thank the DOE/DoD Memorandum of Understanding for funding this research.

7. References

1. Flesner, R.L., Dell'Orco, P.C., Spontarelli, T., Bishop, R.L., Kramer, J.F., Skidmore, C.B., Eaton, E. "Pressurized Alkaline Hydrolysis and Hydrothermal Oxidation of HMX based Energetic Materials as an Integrated Process for EM Disposal." Submitted to *Proceedings of the JANNAF Propulsion and Subcommittee Joint Meetings*, Tampa, Florida, (December 4-8, 1995).
2. Heilmann, H.M., Strenstrom, M.K., Hesselmann, R.P.X., Wiesmann, U., *Wat. Sci. Tech* 30, 53 (1994).
3. Cannizzo, L. F.; Mower, G. L.; Hunstman, L. R.; Achatz, W. R.; Edwards, W. W., *J. Energ. Mater.* 13, 331 (1995).
4. Borcherding, R., "Treatment of Solid Propellant Manufacturing Wastes, Base Hydrolysis as an Alternative to Open Burning," *Proceedings of the JANNAF Safety and Environment Protection and Propellant Development and Characterization Subcommittee Joint Meetings,* San Diego, California, (July 13-14, 1995).
5. Spontarelli, T., Buntain, G. A., Sanchez, J. A., Benziger, T. M., "Destruction of Waste Energetic Materials Using Base Hydrolysis," *Proceedings of the 12th (1993) Incineration Conference,* Knoxville, Tennessee, (May 3rd-7th, 1993) p. 787.
6. Flesner, R. L.; Spontarelli, T.; Dell'Orco, P. C.; Sanchez, J. A. "Base Hydrolysis and Hydrothermal Processing of PBX 9404" Accepted *ACS Symposium Series, Emerging Technologies in Hazardous Waste Management VI.* Sept. 19-21, (1994) Atlanta, GA.

CLEANING OF GASEOUS PRODUCTS FROM THERMAL WASTE TREATMENT

JURGEN VEHLOW
Forschungszentrum Karlsruhe - Technik und Umwelt
Institut für Technische Chemie
Bereich Thermische Abfallbehandlung

1. Introduction

Toxic and dangerous types of waste materials including chemical weapons need to be inertized prior to their final disposal. This can be done by dispersion and encapsulation or solidification of the total waste stream in inert matrices like cement. A different strategy is the separation of critical ingredients by adsorption on adequate substrates like carbon or by absorption in suitable liquids. The most efficient strategy, however, is the complete destruction of all toxic compounds. For inorganic species this requires a respeciation into compounds of acceptable properties. For organic compounds the best strategy in this respect is the total oxidation into carbon dioxide, CO_2, and water, H_2O. This goal may be reached in different ways depending on the methods chosen for destruction. The physical and chemical parameters as well as the temperature of the applied processes can vary in wide ranges.

Biological degradation processes are typically limited to temperatures well below 100° C. They can only be applied to biodegradable species and require sufficient humidity. Thermal treatment processes are distinguished by the applied temperature and the atmosphere in which the reaction takes place. Heating a material in an inert atmosphere causes a decomposition of organic compounds called pyrolysis. Gasification is a conversion of high molecular organics into smaller compounds by means of a reactive gasifying agent (e.g. oxygen or water).

According to this definition combustion is a gasification with air or oxygen as gasifier with a stoichiometry ratio higher than one. Under these conditions oxidation is the prevailing reaction. Combustion of waste makes mainly sense if the heating value allows a self sustained incineration of the material. If the calorific value of a waste stream is too low additional fuel may be added. Such procedure is expensive and can only be recommended if the properties of the waste material require a combustion for total inertization.

The selection of the appropriate process for a defined waste material depends on a number of parameters the most important of which are:

- the chemical nature of the waste,
- its physical properties,
- the mass to be treated,

F.W. Holm (ed.), Effluents from Alternative Demilitarization Technologies, 47–69.

- the environmental impact of the treatment procedure as well as of its products,
- and the economic feasibility of the process.

The decision is, of course, biased by existing legislative regulations. It needs to take into account local conditions including political mainstreams and public acceptance. Hence the decision of the process of choice cannot be made on the basis of scientific, ecological and economical considerations only.

For total inertization in short time and in one single step a direct combustion of the materials in question seems to be the best and economically most promising strategy. The specific properties of various chemical weapons, however, might call for other thermal destruction technologies like pyrolysis or gasification which produce different gaseous products. These gases are still reactive and can be utilized in organic synthesis processes or they need a further treatment before being released into the atmosphere. In most cases this secondary treatment step is a combustion. Such two- or multi-stage processes may allow an easier control of the overall destruction process.

All gases released from waste treatment processes regardless whether they are emitted into the atmosphere or are utilized in further processes need a cleaning prior to their final destination. The applied cleaning strategies depend on the type of the gas and the emission standards or utilization requirements. This paper describes in general terms the most important thermal waste treatment processes and the strategies of gas cleaning associated with those processes. All data are related to the treatment of municipal solid waste. Conclusions into the field of other materials like chemical weapons and their destruction need to consider the differences in the feed stream which will cause treatment products different in quality as well as in quantity.

Most graphs and data used in the following chapters concerning the properties of waste and its incineration are adapted from the handbook 'Municipal Solid Waste Incineration Residues' of the International Ash Working Group [IAWG 1997].

2. Air Emission Standards

The emissions of technical processes into the air are regulated in most countries. The most restrictive emission limits are set for the emissions out of waste combustion plants and it is a fair statement that the respective limits follow the same standards in most industrialized countries. This is evident from the air emission standards of selected countries compiled in Table 1. These emission standards will be used as guideline in the following discussion of cleaning strategies for the offgases of different thermal waste treatment technologies.

TABLE 1: Comparison of mission limits in mg/m^3 (PCDD/F in ng(I-TEQ)/m^3), daily means of dry gases at 273 K, 101.3 KPa and standardized to the O_2 concentration (* : Cd+Tl)

	EU 94/67/EEC	Germany	The Netherlands	USA (MACT)	Canada (Ontario)	Japan (act. criteria)
O_2 (%)	11	11	11	7	11	12
Dust	10	10	5	24	17	10 - 50
HCl	10	10	10	25	18	15 - 50
HF	1	1	1	-	-	-
SO_2	50	50	40	30	21	10 - 30
NO_x	-	200	70	150	110	30 - 125
CO	50	50	50	100	-	50
Hg	0.05	0.05	0.05	0.08	0.057	0.03 - 0.05
Cd	0.05*	0.05*	0.05	0.02	0.014	-
TOC	10	10	10	-	-	-
PCDD/F	0.1	0.1	0.1	0.14 - 0.21	0.14	0.5

3. Composition of Municipal Solid Waste

Municipal solid waste is a composite of more or less all chemical elements each present in different species. This includes even hazardous pollutants such as dibenzo-p-dioxins (PCDD) and dibenzofurans (PCDF). The most abundant element is the carbon with approx. 30 wt. %. In combination with approx. 4 wt. % of hydrogen bound to organic compounds it controls the calorific value. The today's typical lower heating value of municipal solid waste in industrialized countries is in the order of 8 - 10 with maximum values up to 14 MJ/Kg. Ingredients reducing the heating value are the incombustibles (approx. 20 - 40 %) and especially the humidity of typically some 25 - 35 %. The latter one is mainly originated from the green and kitchen waste fraction. The water content can vary to a great extent depending on season and weather conditions. In developing countries with a higher fraction of putrescibles in the waste stream up to 70 % of moisture have sometimes been found.

Elements of concern in the context of this paper are those which are characterized by a high toxicity and a high volatility under the conditions of waste combustion. The typical concentration ranges of a number of such volatile toxic elements and of the major components in the waste of industrialized countries is compiled in Table 2 [Johnke 1994, IAWG 1997]. With the only exception of As all volatile elements in the right part of the table show a significant enrichment compared to their occurrence in the lithosphere and hence have to be controlled in any thermal waste treatment process.

TABLE 2: Concentration ranges of selected elements in municipal solid waste [Johnke 1994, IAWG 1997]

Element	Concentration mg/Kg		Element	Concentration mg/Kg
C	250 000 - 400 000		As	3 - 10
H	50 000 - 60 000		Cd	5 - 15
N	5 000 - 10 000		Sb	10 - 60
Cl	5 000 - 8 000		Hg	0,5 - 5
Br	30 - 200		Tl	0.1 - 1
S	1 000 - 3 000		Pb	400 - 1 000
Se	0.5 - 1		PCDD/F	0.000 02 - 0.0001

The concentration of PCDD/F in the waste of 20 - 100 mg/Mg is by a factor of approx. 20 higher than the today's value found in so-called 'uncontaminated' soils in Central Europe.

4. Processes for Thermal Waste Treatment

4.1. CLASSIFICATION

Thermal treatment of waste can be performed at different temperatures using different process conditions. The various processes are schematically compiled in Table 3 with the most commonly applied ones for waste treatment being marked.

TABLE 3 Thermal waste treatment options

Temperature range in °C	Inert atmosphere	Controlled addition of gasifier	oxygen surplus
< 150	drying	drying	drying
<500	low-T degassing		
500 - 800	**pyrolysis**	**gasification**	
>800	high-T pyrolysis		**Combustion**

In the following the fundamental processes combustion, gasification and pyrolysis will be described. This includes the major process techniques and the products with special emphasis devoted to the gaseous products.

4.2. COMBUSTION

4.2.1. Combustion Principles

The most widely applied option for thermal waste inertization is the combustion which is the total oxidation of the waste stream in an oxygen surplus atmosphere. Waste combustion is very common in most industrialized areas and countries like Switzerland, Denmark or Japan incinerate up to 80 % of their municipal solid waste. A typical system for waste combustion is the European mass burner, a facility which incinerates the waste on a grate without prior treatment.

The generic physico-chemical processes in the combustion chamber of such system are depicted in Fig. 1. The waste is typically fed in without any pretreatment. The radiation of the hot walls cares for a drying of the material at the front end of the grate. At increasing temperature pyrolysis and gasification processes cause the fragmentation of organic compounds and their major transfer into the gas phase. The combustion of these products in the gas phase contributes up to 80 % to the total energy release. At further increased bed temperatures the remaining skeleton or fixed carbon is combusted. The back-end of the grate, the so-called burnout region, is intended to care for a complete conversion of residual carbon and a sintering of the bed material.

A number of different grate systems has been developed (e.g. reciprocating, reverse acting, roller grates) which guarantee an excellent burnout in all residue streams including the gas phase.

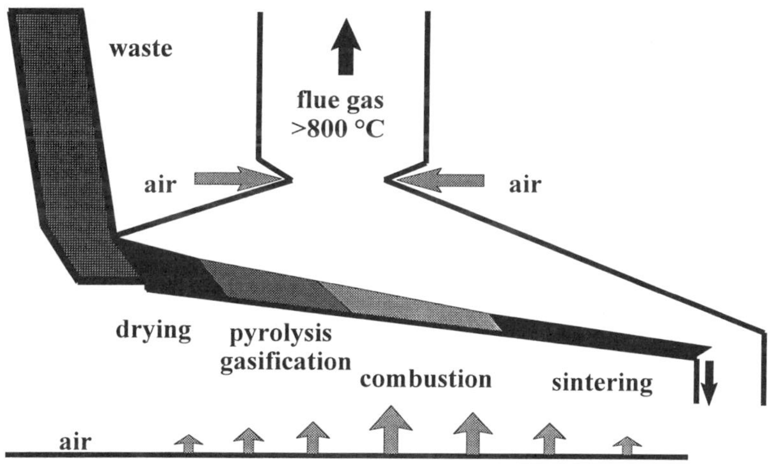

Figure 1. Processes inside the combustion chamber of a grate system

Especially in Japan a different type, the fluidized bed incinerator, is often used for waste combustion. This type of combustion device calls in most cases for a size reduction of the waste. Hazardous waste requiring high combustion temperatures is mainly burnt in a rotary kiln which is followed by a post-combustion chamber. A rotary kiln can treat solid, liquid and gaseous waste materials. Most waste combustion plants in chemical industry use this configuration. In the few cases where only liquid or gaseous fractions have to be treated, a combustion chamber equipped with an adequate burner can be used.

The combustion of waste is, regardless of the combustion technology, controlled by the same chemical reactions and thermodynamic equilibria. The kinetics differ in different furnaces and can influence the actual generation of pollutants as in the case of the conversion of nitrogen compounds to nitrogen oxides.

4.2.2. Combustion products

The high temperature and the oxygen surplus in a combustion chamber care for a total conversion of all input species into the most stable products under oxidizing conditions. Furthermore a partitioning between species staying in the fuel bed and those entering the gas phase takes place. The major primary combustion products are shown in Fig. 2. Mineral materials as well as lithophilic heavy metals like iron, manganese, titanium, nickel or copper and their compounds leave the grate as bottom ashes at the back-end. A certain amount of salts like alkali or earth-alkali chlorides and especially sulfates stays in this residue, too.

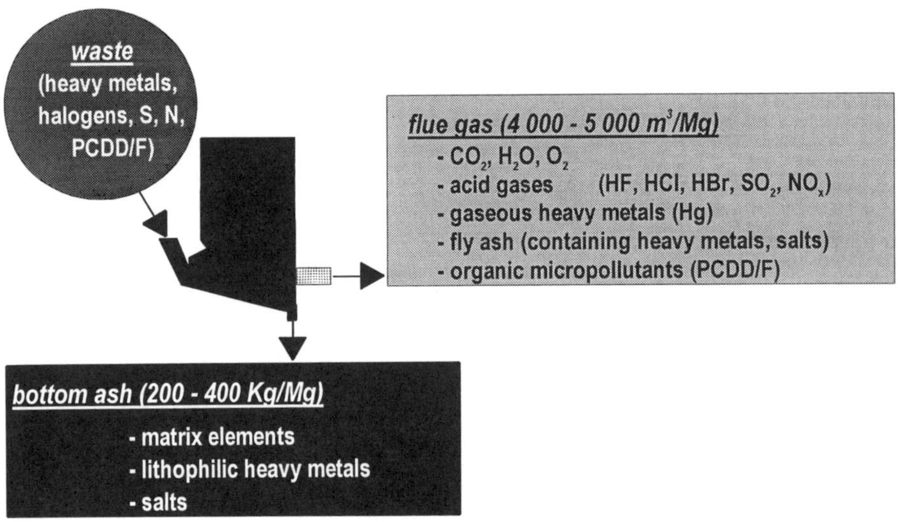

Figure 2. Partitioning in the furnace of a waste combustion facility.

The stable oxidation product of all carbon compounds is CO_2. Hydrogen is always oxidized to H_2O. Both enter the gas phase. Sulphur may form the gaseous SO_2 or sulfates which stay in the bottom ashes. The waste bound nitrogen is partly oxidized to NO with small amounts of other oxides present like NO_2 and N_2O. The oxidation of air nitrogen has no importance since the combustion temperature is too low. Depending on the operation mode more or less particulate matter is transferred out of the fuel bed and leaves the furnace along with the flue gas.

Halogens are transformed into the respective halides. In the case of Cl the major product is the gaseous HCl. Only 10 - 15 % of the inventory stays as chloride in the solid residues. For Br the partitioning between HBr and bromides is about equal and for F approx. 70 - 80 % are found as fluorides in the solids.

Volatile heavy metals are mainly evaporated as chlorides in the furnace. More than 80 % of the inventory of Hg, Tl and Cd leave the combustion chamber along with the flue gas. Zn, As, Sb and Pb show an approximately even partitioning between bottom ash and raw gas. At the typical boiler outlet temperature of 180 - 200 °C, however, all volatile heavy metal species but Hg and its compounds are bound to particulate matter and are discharged along with the filter ashes.

The oxidation of the carbon inventory of the waste results not only in CO_2 but minor amounts of products of incomplete combustion such as CO, PAH, or particulate carbon are formed, too. The latter one is known to be a source of toxic low-volatile organic micro-pollutants, especially of PCDD and PCDF. Their major formation route is an oxichlorination of reactive particulate carbon in the temperature range 200 - 500 °C inside the boiler. The reaction is catalyzed by Cu [Vogg 1986, Hagenmaier 1987, Hunsinger 1995]. Some 10 years ago PCDD/F raw gas concentrations downstream of the boiler were in the order of 10 - 100 ng(I/TEQ)/m^3. The improvement of the burnout caused a drastic reduction of this level. A today's state-of-the-art waste combustion facility has raw gas concentrations <1 ng(I/TEQ)/m^3 [Reeck 1991].

Typical concentrations and mass flows of some components in the raw gas of a municipal solid waste combustion facility are put together in Table 4.

TABLE 4. Ranges of raw gas concentrations and averaged mass flows in a municipal solid waste combustion plant (*: ng(I-TEQ)/m^3, **: mg(I-TEQ)/Mg)

	concentration (mg/m^3)	mass flow (Kg/Mg)		Concentration (mg/m^3)	mass flow (Kg/Mg)
O_2	60 000 - 120 000	500	NO	100 - 500	2
CO_2	150 000 - 200 000	900	NH_3	5 - 30	0.1
H_2O	110 000 - 150 000	600	CO	<10 - 30	0.1
dust	1 000 - 5 000	20	TOC	1 - 10	0.02
HCl	500 - 2 000	6.5	Hg	0.1 - 0.5	0.002
SO_2	150 - 400	2	PCDD/F	0.5 - 5*	<5**

These raw gas concentrations have to be reduced by an adequate air pollution control system in order to comply with the respective legislative emission standards before released into the air.

4.2.3. Air Pollution Control Systems

4.2.3.1. Principles The abatement of pollutants from their raw gas levels down to concentrations well below the respective air emission standards can be achieved in many ways. The design of the various configurations found in full scale plants depends not that much upon the clean gas quality but on investment and/or operation cost, utilization or disposal option of the residues or on available space in the case of upgrading of old facilities.

It is obvious that more than one single purification stage is needed to reduce the raw gas concentrations of the pollutants compiled in Table 4 down to, or even below, the emission limits. According to their chemical properties different types of control technologies have to be implemented which care for:

- precipitation of particulate matter,
- removal of acid gases by liquid or solid absorption respectively neutralization,
- removal of Hg (and eventually other gaseous metals) by absorption or adsorption,
- catalytic or non-catalytic reduction of nitrogen oxides,

- adsorption of organic compounds or their destruction by oxidative, catalytic or irradiation measures.

Today all technologies and all kinds of combinations of abatement options can be found in full scale installations. Each configuration guarantees the compliance with the today's most stringent air emission standards. A discrimination can only be made on the basis of economic aspects and the amount and/or hazardousness of the residues.

4.2.3.2. Dedusting. In most plants the first gas cleaning step in an air pollution control (APC) system is the dedusting. Due to its simple design and easy operation electrostatic precipitators (ESP) are often used for this purpose. A modern ESP guarantees dust emissions of only few mg/m^3. Even lower values can be achieved with fabric filters. This type is preferred in dry systems. The material used in fabric filters limits the flue gas temperature to 200 - 220 °C. The risk of ignition by hot particles is another reason why fabric filters are rarely found as dedusting devices directly downstream of the boiler.

An excellent dedusting down to the mentioned order of magnitude guarantees the compliance with the emission standards of heavy metals except of Hg. The particle bond PCDD/PCDF are precipitated, too. This, however, is only helpful in old facilities. Modern municipal solid waste combustion plants which are characterized by a substantial amount of PCDD/PCDF being transported in the gas phase need additional abatement strategies.

Some efforts have been made to implement hot gas filtration at temperatures above 500 - 600 °C in waste combustion plants in order to minimize the formation of PCDD/F by removing the reactive ingredients before the gas temperature reaches the formation window of these compounds. In that temperature range there is a risk of clogging of the ceramic filter candles by sticky dust particles and an effective cleaning is extremely difficult. Similar experiences have been made with the cleaning of high temperature cyclones which have the additional disadvantage of a low removal efficiency for fine particulate matter.

Hence the benefits of high temperature dedusting are more or less compensated by technological complications and this technology is not used in waste combustion today. In other processes like gasification where hot gases are directly operating a turbine this technology is of great importance.

4.2.3.3. Scrubbing Technologies for Acid Components. The following step in the air pollution control system is chemical gas cleaning, which can be designed in four different ways: as a wet, semi-wet, semi-dry or dry system. A common scrubbing strategy in Central Europe is the wet APC system, today in most cases performed as a two-stage installation with an initial acid scrubber followed by a neutral/weakly alkaline scrubber. Such two-stage systems are very efficient in removing halogen hydrides typically down to 1 - 5 mg/m^3 for HCl and < 1 mg/m^3 for HF and HBr. If the pH is kept low (< 1) Hg emission levels far less than 50 mg/m^3 can easily be achieved [Braun 1992]. The second scrubber reduces SO_2 down to the order of 10 mg/m^3.

Various designs of wet scrubbers are in use with each having special advantages and disadvantages. E.g. jet scrubbers have a high dust removal capability which has to be paid for by a high pressure drop and that means energy consumption. A number of common wet scrubber configurations is shown in Figure 3.

Wet systems are operated close to neutralization stoichiometry and are hence producing a minimum amount of (salt) residues. The neutralizing agent may be NaOH or $Ca(OH)_2$. Additives are often used to remove traces of heavy metals like Hg or Cd. In some plants NaOH is used for the first scrubber and $Ca(OH)_2$ for the second one in order to recover the sulphur as gypsum.

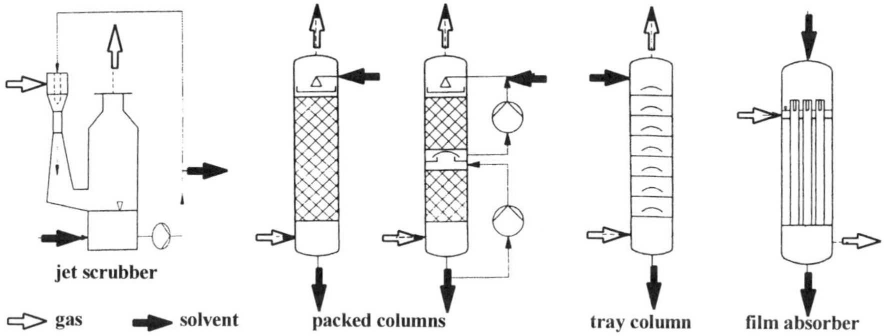

Figure 3. Various designs of wet scrubbers

Figure 4 shows the design of an APC using wet scrubbing with neutralization/cleaning and discharge of the effluents to a sewer. An estimate of the residues produced by this configuration is included. The scrubbing residue is calculated as dry sodium salt. The filter ashes as well as the neutralization sludges have to be disposed of on special disposal sites due to their high content of various pollutants like heavy metals and PCDD/F.

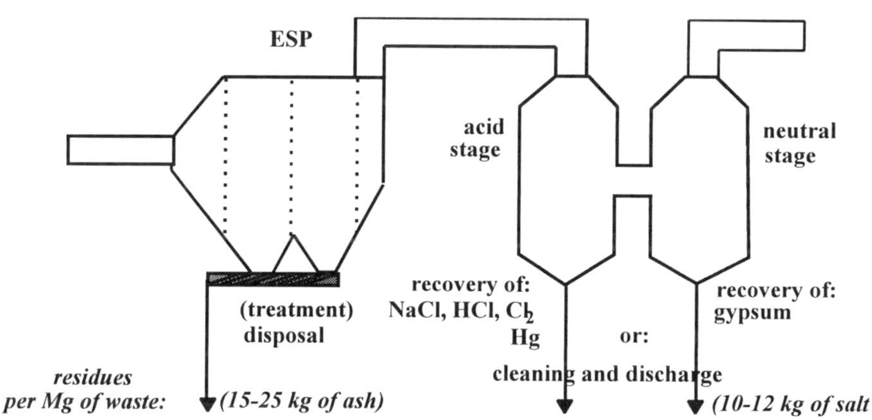

Figure 4. APC system with wet scrubbing (discharge of effluents)

If an efficient dedusting can be achieved the wet scrubbing system allows the recovery of NaCl [Jekel 1992], HCl [Kürzinger 1989] or elementary chlorine [Volkman

1991] from the acid scrubbing solution and of gypsum from the neutral scrubber's effluents [Kürzinger 1989]. The respective recovery options are noted in Figure 4. Few of these processes have been realized in full scale plants due to limited markets for the products or due to the economy of the recovery process.

In some Central European countries, first of all in Germany, authorities try to ban the discharge of liquid effluents. Hence the scrubbing solutions need to be evaporated after neutralization. The easiest way to do this is by means of a spray drier installed upstream of the scrubbing system. In this case a primary fly ash removal may not be economical and the dust is discharged along with the salts in the (fabric) filter downstream of the spray drier. A typical configuration is depicted in Fig. 5. This type of wet scrubbing leaves the operator with the problem where to dispose of the highly soluble salt residue the amount of which is - depending on the neutralizing agent - of the same order as in the operation mode with water discharge.

The other abatement strategies mentioned above, the semi-wet, semi-dry and dry systems, use the reactivity of injected ground lime (dry), dry $Ca(OH)_2$ with prior humidification of the flue gas (semi-dry) or $Ca(OH)_2$ slurry (semi-wet) to trap the acid components of the flue gas. This can be done with or without an initial fly ash removal.

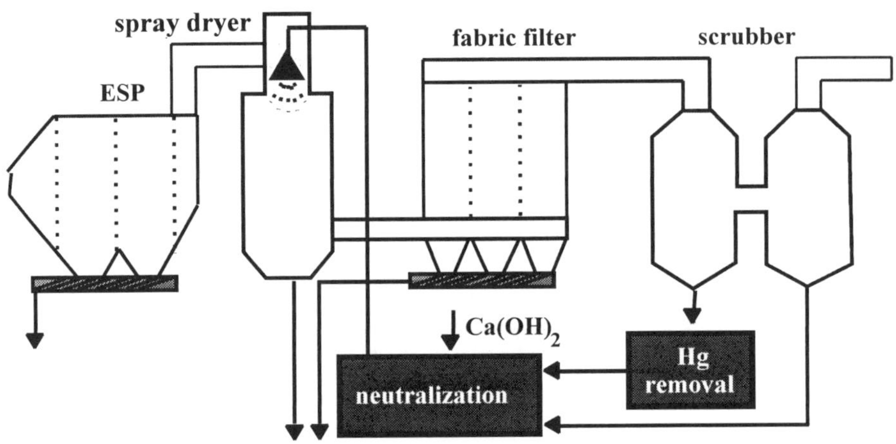

Figure 5. APC system with wet scrubbing and evaporation of neutralized scrubbing solutions

The basic design of the systems in question is always similar to that given in Figure 6. Their advantage is their simple design and operation. Their disadvantage compared to the wet system is the higher stoichiometry albeit in most installations the reagent is recycled to a certain extent. The stoichiometric factor varies between 1.8 for the semi-wet up to >3 for dry systems. The reagent surplus causes higher amounts of residues which increase in the above given order of the methods and hence cause higher disposal costs.

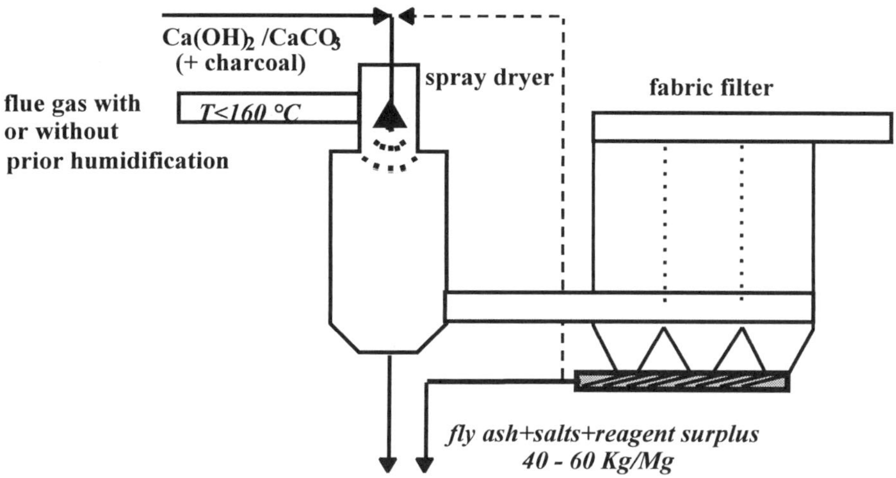

Figure 6. APC system with dry scrubbing

A different dry scrubbing technology that is mainly promoted in France uses hydrogencarbonate for the removal of acid components. The process enables a recycling of components [Korte 1994].

4.2.3.4 Removal of NO_x. A separate problem in all systems is the NO_x abatement which is required in some European countries and will most likely be part of the future European Union's regulation. This gas is either reduced using selective non-catalytic reduction (SNCR) at temperatures of approx. 900 °C by injecting ammonia or an amine into the raw gas or by means of selective catalytic reduction (SCR) at the back-end of the APC system at temperatures of 300 - 350 °C. The disadvantage of the latter process is the need for heating the flue gas up, its advantage is the higher efficiency compared to the SNCR process.

This problem can be overcome if the catalyst is placed downstream of the filter. This means, however, that the filter has to be operated at high temperature with the risk of additional PCDD/F breeding (see below) or that a low-temperature catalyst needs to be installed which is yet not state of the art.

A different approach to reduce NO_x is the electron beam irradiation [Paur 1993]. The process is also reducing the SO_2 level [Mätzing 1993]. It has been tested in larger scale in coal fired power plants. A first implementation in a waste combustion plant has been reported from Japan.

4.2.3.5. Removal of PCDD/F. Another group of pollutants, the low-volatile polychlorinated organic compounds like PCDD and PCDF require special attendance, too. These compounds are easily adsorbed on carbon surfaces. Hence in old waste combustion plants with concentrations of particulate carbon in the fly ashes in the order of several percent the PCDD/F could efficiently be removed by good dedusting. In modern installations the low PCDD/F raw gas levels reached by an improved burnout go along with low concentrations in particulate C, too, and a great fraction of the

PCDD/F stays in the gas phase. These gaseous species are not affected by the above described scrubbing technologies and have to be taken care of by special measures.

Gaseous PCDD/F can be adsorbed on charcoal in a bed filter [Dannecker 1990] or in an injection system [Nielsen 1989]. For smaller plants with flue gas volumes of < 50 000 m^3/h the adsorption of PCDD/F on plastics at temperatures < 100 °C has been proposed [Kreisz 1995]. This adsorber allows an easy desorption of the PCDD/F at temperatures of 130 °C and their thermal destruction in the combustion chamber. If a catalytic Denox system is installed it can be modified in a way that the back-end of the catalyst is operated at oxidizing conditions for the destruction of PCDD/F [Hiraoka 1989]. The electron beam irradiation mentioned above as an alternative Denox method has also the capability to destroy PCDD/F [Paur 1992].

All of these abatement techniques are installed downstream of the scrubber and in the case of catalytic destruction a heating of the flue gas is required. An intervention early in the process chain is the destruction by injecting H_2O_2 as oxidizing agent into the raw gas between boiler and filter [Vogg 1990]. This technology is most promising for wet systems since it prevents the organic compounds from entering and polluting the scrubbers [Kreisz 1985]. The process requires a fabric filter for dedusting.

In dry, semi-dry or semi-wet systems small amounts of charcoal can be injected upstream of the fabric filter along with the neutralizing agent. This guarantees an efficient removal of Hg, too.

4.2.3.6 Environmental Relevance of Emissions. About ten years ago growing opposition against waste combustion and the desire to improve its public acceptance were the driving force for an almost unlimited extension of APC systems in Germany. One example of a so-called 'advanced' configuration is shown in Fig. 7. This development was accompanied by an extreme increase in combustion costs. For the time being the tendency can be observed to focus more interest onto the improvement of primary measures like combustion control in order to simplify the APC system [Vehlow 1996].

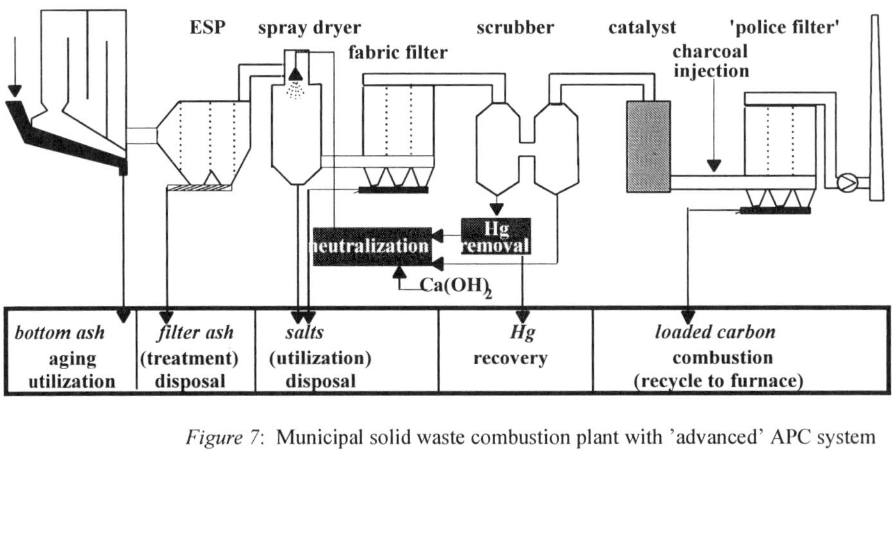

Figure 7: Municipal solid waste combustion plant with 'advanced' APC system

To speculate about future developments it might be useful to scrutinize the environmental effects of the emissions of a modern municipal solid waste incinerator. This has been done in Table 5 for a selection of components. The second column presents typical emission levels out of state-of-the-art facilities. The third column shows mean concentrations of the components in the ambient air [Vogg 1992] and the last column gives additional emissions calculated from the emission values by means of a dispersion model assuming a volume stream of 100 000 m^3/h and a stack height of 65 m [SCAQMD 1991].

TABLE 5: Environmental relevance of the stack emissions from waste combustion

Component	background concentration	stack emission	additional emission
HCl	30 µg/m³	5 mg/m³	0.15 µg/m³
SO_2	20 µg/m³	20 mg/m³	0.6 µg/m³
Dust	40 µg/m³	1 mg/m³	0.03 µg/m³
Cd	3 ng/m³	5 mg/m³	0.15 ng/m³
Hg	5 ng/m³	10 mg/m³	0.3 ng/m³
PCDD/F (TEQ)	100 fg/m³	0.05 ng/m³	1.5 fg/m³

The table furnishes evidence that the stack emissions of a modern waste combustion plant give rise to only minimal increases in the low percent level of the already existing background concentrations. These additional emissions do not cause any environmental risk. This is even valid for the critical components PCDD/PCDF and Hg. Hence further reduction of the emission limits as well as further efforts to improve the efficiency of flue gas cleaning systems cannot be justified by an improvement of the environmental compatibility of waste incineration.

4.3. GASIFICATION

Gasification by means of oxygen has long been used to generate synthesis gas from lignite or hard coal. During the last decades attempts have been made to modify this process for the treatment of waste or waste fractions like plastics. The idea is to convert organic waste ingredients into a synthesis gas which can either be used for chemical synthesis, e.g. for methanol, or for energy generation in a gas turbine, a gas motor or in a combustion chamber. The composition of the synthesis gas depends on the gasifying conditions and the temperature. The major gaseous products of the process are carbon monoxide, hydrogen and short chain aliphatic compounds. Gasification is an endothermic process. The energy to maintain the process is often generated by partial combustion. Hence CO_2 is another major product of most proposed gasification processes.

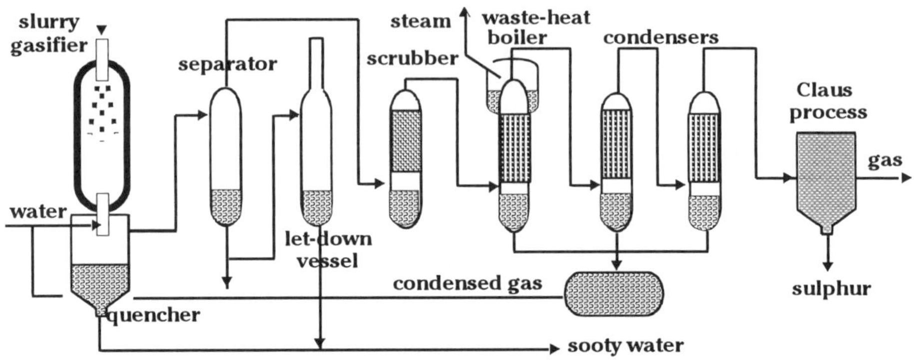

Figure 8: Scheme of the LAUBAG entrained-flow gasification process [Buttker 1993]

The LAUBAG in Germany operates pressurized fixed-bed and entrained-flow systems for the gasification of lignite which have been modified to treat up to 50 % of ground plastic waste [Buttker 1993]. The final product is methanol which means that the synthesis gas requires some efforts with respect to adequate purification. A draft scheme of the entrained-flow gasification process is shown in Fig. 8.

The gasification agent is a steam-oxygen mixture. The hot crude gas is quenched with water which causes the formation of substantial amounts of soot. The quench water contains also ashes, phenols and a tar/oil mixture the separation of which is complicated (and expensive). The gas undergoes several condensation and purification processes. The gasification product of nitrogen compounds, NH_3, is said to be neutralized by the HCl in the raw gas and removed in the scrubber which is neutralized with NaOH. The major sulfur product H_2S is converted into elementary sulfur by means of a modified Claus process, a catalytic oxidation according to the chemical equation: $2\ H_2S + SO_2\quad 3S + 2\ H_2O$. Regarding this gas cleaning chain it becomes evident that the presence of a wide range of different organic and inorganic gaseous compounds complicates the gas purification procedure significantly.

A self-sustained gasification by partial oxidation requires an excellent control of the oxygen supply to establish defined partial oxidizing conditions in an overall reducing atmosphere. A hindrance in that respect is the inhomogeneity of the waste. This can be overcome by a size reduction (as in the case of the entrained-flow gasification discussed above) or by other pretreatment measures. One option is a degassing/pyrolysis as a conditioning stage prior to a high temperature gasification.

Such combinations have been developed by Noell and Thermoselect. The Noell Conversion Process is a combination of a pyrolysis in a rotary drum and an entrained-flow gasification which has been tested in pilot scale [Schingnitz 1997]. The Thermoselect process is characterized by a direct coupling of pyrolysis and gasification and will be described more in detail [Stahlberg 1995, Häßler 1995].

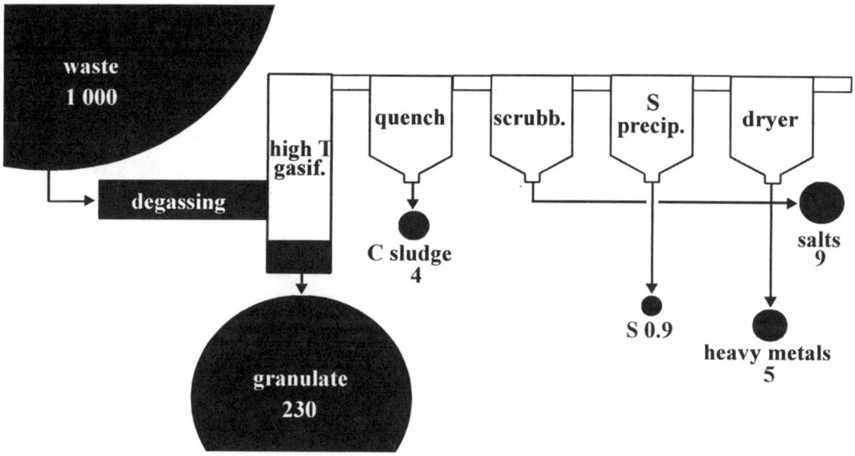

Figure 9: Scheme and mass balance of a Thermoselect plant (data in Kg/Mg of waste)

A scheme of the Thermoselect process is shown in Fig. 9. Untreated waste enters a channel where it is compacted and heated up to maximum temperatures of approx. 600 °C. In this channel a degassing takes place. The gases and the heated and partially pyrolyzed solids enter a chamber where they are gasified by injection of pure oxygen. The controlled combustion of parts of the material supplies the energy for the overall process.

The solids are molten at the bottom of the chamber at temperatures of approx. 2 000 °C resulting in approx. 230 Kg of silicate and oxide matter including a certain amount of an iron alloy. The solids are discharged as granulate via a quench bath.

The temperature in the gas phase is kept at about 1 200 °C. Approx. 1 000 m^3 of synthesis gas are produced per Mg of gasified waste which are directly quenched by water injection down to less than 90 °C. The range of the concentration of the major constituents as well as the averaged mass flows in the dry quenched gas are compiled in Table 6.

It is evident that the gas quality is rather different from that in combustion plants. The products for synthesis or energy production are CO and H_2. Due to the high temperature there are only low concentrations of organic compounds present. Components of concern are some reduced species like H_2S, HCN or NH_3 which have to be removed in the gas cleaning stage. The applied methods are quite different from those known from combustion gases. The scheme of the gas cleaning strategy of a Thermoselect plant is given in Fig. 10.

TABLE 6: Ranges of concentrations and averaged mass flows in the quenched gas in a Thermoselect plant downstream of the quencher (*: ng(I-TEQ)/m^3, **: mg(I-TEQ)/Mg)

	concentration (mg/m^3)	mass flow (Kg/Mg)		concentration (mg/m^3)	mass flow (Kg/Mg)
O_2	<1 000	<1	SO_2	800 - 3 000	1.5
CO_2	400 000 - 500 000	410	HCN	1 - 10	<0.01
CO	350 000 - 400 000	420	N_2	20 000 - 25 000	25
H_2	22 000 - 30 000	27	CH_4	<0.5	<0.0005
HCl	100 - 300	0.2	C_nH_m	<0.1	<0.0001
H_2S	600 - 1 500	0.9	PCDD/F	<0.05*	<0.05**

The graph reveals a rather complex gas treatment system with many stages for the gas as well as for the liquid phase. The direct quenching of the hot synthesis gas causes the formation of approx. 4 Kg of soot and fly ashes per Mg of waste. This soot is filtered out of the quench solution and is ought to be recycled to the degassing channel. Whether this will be practised in a full-scale plant has to be waited for, since there is a risk of creating unwanted internal cycles of volatile heavy metals. The liquid undergoes a neutralization and a sulfide precipitation for heavy metal removal. The process ends in the production of a mixed salt.

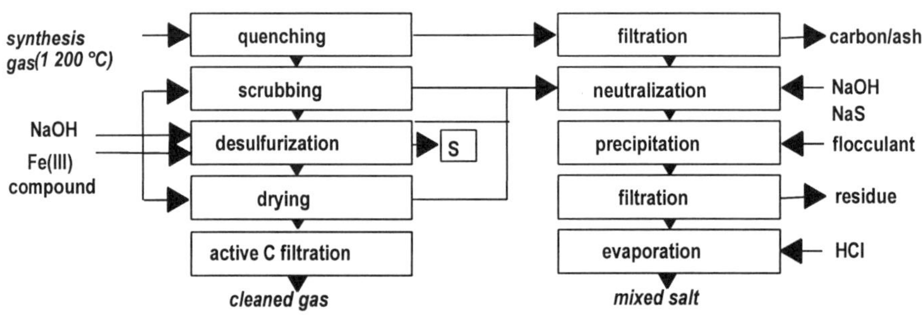

Figure 10: Cleaning of the synthesis gas in the Thermoselect process

The quenched synthesis gas enters the scrubbing stage without any further dedusting. The scrubbing starts with a removal of acid components like HCl in a scrubber neutralized with NaOH. A process for the oxidation of H_2S by means of an Fe(III)-compound, the Sulferox scrubbing, has been integrated which results in the production of elementary sulfur. The primary product, however, contains only 40 % of sulfur and has to be cleaned. Whether the final products will find a market has to be shown. There is reason to speculate about future simplification of the gas cleaning procedure in order to reduce the costs.

Thermoselect operates a pilot plant in Northern Italy and is building up a full scale plant in Karlsruhe, Germany, with an annual capacity of 225 000 Mg realized in three 10 Mg/h lines.

4.4. PYROLYSIS

Pyrolysis is the decomposition of organic compounds at high temperatures in an inert atmosphere. Depending on the input material and on the temperature different reactions prevail and different pyrolysis products have to be expected. A selection of important reactions is reported in Table 7.

A number of pyrolysis processes has been developed for municipal solid waste or for special waste fractions such as used tires or mixed post consumer plastics. The reaction devices were mainly rotary drums or fluidized beds, but other types of reactors like pit, cupola and chamber furnaces or steam crackers have been tested, too. As can be deduced from Table 7, the pyrolysis temperature has a great influence on the partitioning of the pyrolysis products between solid, liquid and gas phase.

TABLE 7: Typical pyrolysis reactions prevailing in the respective temperature ranges [Lenz 1979]

T in °C	(chemical) reactions
100 - 120	Drying
250	split off: H_2O, CO_2, H_2S, HCl
340	destruction of aliphatic bonds, split off: CH_4, aliphatics
380	formation of pyrolysis coke
400	splitting of C-O and C-N bonds
400 - 600	transformation of bitumen to pyrolysis oil and tar
600	cracking of long-chain compounds, formation of aromatics
>600	formation of butadien [3] cyclohexane, benzene, PAH

The temperature influence on the pyrolysis of municipal solid waste has been tested in laboratory scale with the results depicted in Fig. 11. The left graph shows the distribution of the pyrolysis products on the different phases. The most important consequence is the increased formation of gaseous products with increasing temperature. The composition of the products in the single phases is controlled by the pyrolysis temperature, too. The right graph in Fig. 11 indicates that in the gas phase at high temperature hydrogen and CO prevail. The formation of methane has a maximum efficiency in the range of 600 - 700 °C.

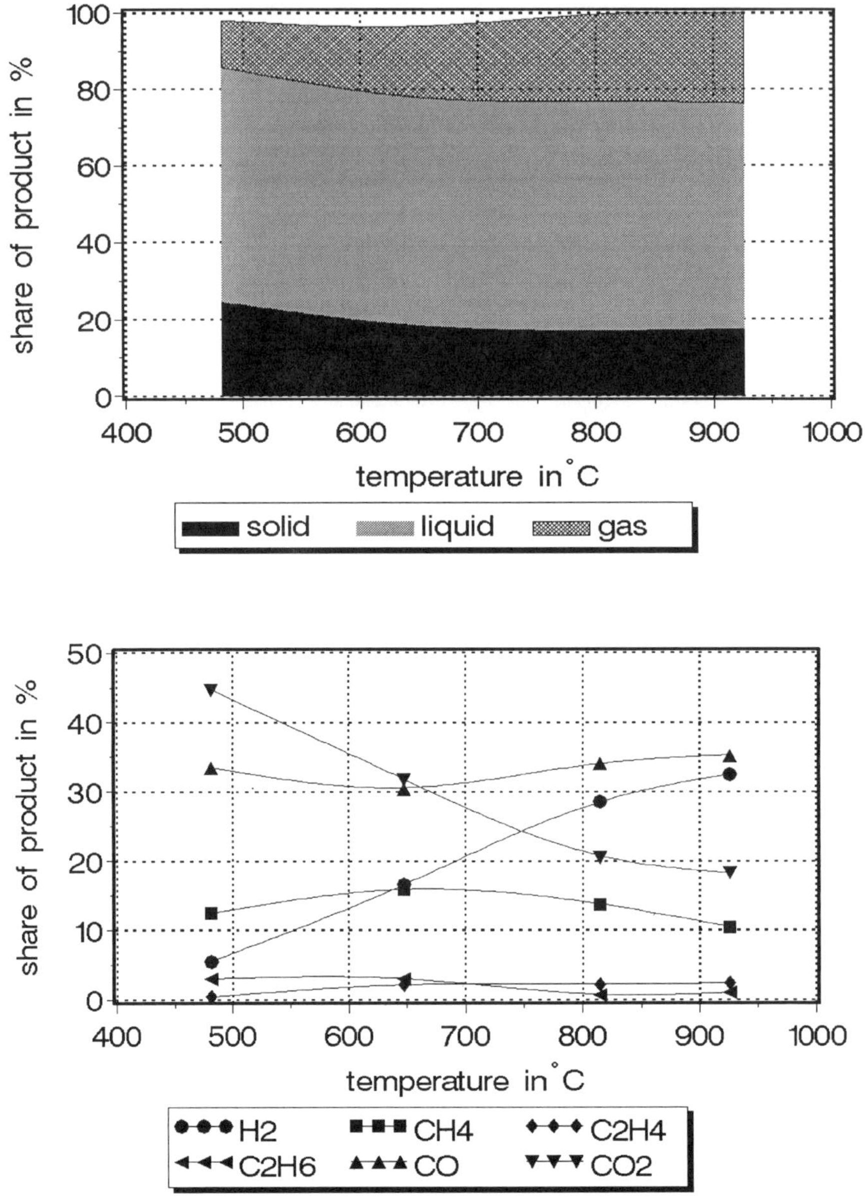

Figure 11: Partitioning of pyrolysis products (left graph) and composition of the gas phase (right graph) as a function of temperature [Tchobanoglons 1977, Sattler 1982]

The spectrum of the pyrolysis products changes significantly if the input material is changed. Hence a number of pyrolysis processes has been proposed and sometimes tested in laboratory scale for specific waste fractions. E.g. the oil crisis promoted attempts to modify the pyrolysis of tires and plastics towards a high output of gasoline like organic compounds [Kaminsky 1979]. A fine tuning of the conditions could shift the pyrolytic reaction to the formation of benzene and its homologues. These products are also useful for specific chemical syntheses. The control of input quality and pyrolysis parameters as well as the complex procedures for the isolation of the required compounds, however, are endangering the economy of such processes.

Most published results from pyrolysis experiments do not show any inorganic products in the gas phase. The formation of acid gases can to a great extent be suppressed by addition of lime or other alkaline material to the feed stream. Nitrogen, sulfur (especially if tires are pyrolyzed), phosphorus, selenium and other elements may form hydrogen compounds which are often toxic and odorous and have to be removed from the gas stream prior to any utilization.

Some results from laboratory and full scale on the pyrolysis of waste (mainly household waste, but also sewage sludge and other waste fraction) at 600 - 650 °C have been compiled in Table 8. The data are rough estimates of the ranges to be expected only. They indicate, however, that the cleaning of the pyrolysis gas has to treat a number of ingredients with a rather wide spread of chemical properties.

TABLE 8: Ranges of concentrations and averaged mass flows in the humid pyrolysis gas from municipal solid waste at 600 - 650 °C [Bilitewski 1985, Fichtel 1987]

	concentration (mg/m^3)	mass flow (Kg/Mg)		concentration (mg/m^3)	Mass flow (Kg/Mg)
CO_2	150 000 - 250 000	90	NH_3	20 - 100	0.03
CO	70 000 - 120 000	45	HCN	1 - 50	0.01
H_2	4 000 - 6 000	2.5	CH_4	30 000 - 40 000	18
SO_2	<100	<0.05	C_2H_4	10 000 - 15 000	6
H_2S	>1 000	>0.5	C_2H_6	14 000 - 18 000	8
COS	traces		C_nH_m	traces	

For the time being such processes are not competitive on the market due to the high expenses for product cleaning. The logical consequence is the direct combustion of the pyrolysis products. This concept is mainly followed in areas where the major aims are either soil remediation or the pyrolysis of residues from the electrical and electronic sector where the recovery of valuable metals was the major focus.

On this basis two different concepts have been realized in full scale in Germany during the last decade. The first pyrolysis plant has been built up by Deutsche Babcock Anlagen in 1987 in Günzburg [Comfére 1985, Fichtel 1987]. It was designed for an annual throughput of 24 500 Mg of municipal solid waste, 6 000 Mg of waste from trade and light industry and 4 000 Mg of dry sewage sludge. The waste is pyrolyzed at 600 °C in two indirectly heated rotary drums. The pyrolysis gas is dedusted by cyclones and burnt at 1 200 °C in one high temperature combustion chamber. The energy is recovered in a boiler. The addition of lime to the waste input reduces the levels of acid species in the raw gas. Hence a simple semi-dry gas cleaning system is installed which is operated with an addition of activated coke in order to guarantee the effective removal of PCDD/F and Hg.

The weak point of this concept is the solid residue which is correctly addressed as pyrolysis coke. The disposal of such materials with high carbon contents will be banned in Germany and other countries in the first years of the next century. This problem has been solved in the Siemens Thermal Waste Recycling Process [Berwein 1990, Anton 1995] a scheme of which is shown in Fig. 12.

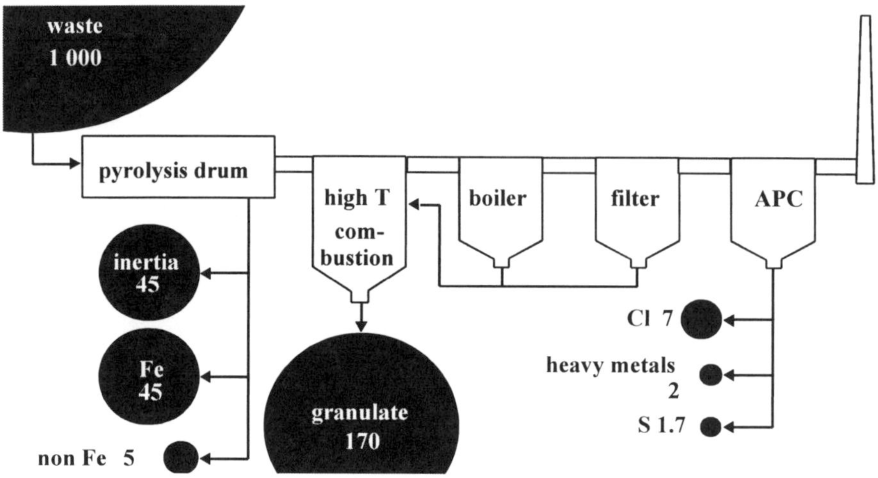

Figure 12: Schematic drawing and mass balance of a Siemens Thermal Waste Recycling Plant (data in Kg/Mg of waste)

The first stage is a pyrolysis of shredded waste in an indirectly heated rotary drum. The carbon fraction in the solid residues is segregated and after grinding burnt together with the pyrolysis gases in a high temperature (approx. 1 350 °C) combustion chamber. The gas cleaning is conventional as in waste combustion plants and starts with an ESP. The original layout contains a total recycling of the boiler and filter ashes back into the high temperature combustion. To prevent the establishment of internal cycles of high volatile metals like Cd this practise has been changed and an unknown part of the ashes is discharged today.

The ESP is followed by a two-stage wet scrubbing system. This system is intended to allow the production of HCl and gypsum. The amount of 2 Kg of heavy metals separated from the effluents documents that the dedusting is not very efficient and the question is allowed, whether a production of marketable materials from such contaminated solutions will be economically feasible.

The process has been tested for a number of years in a pilot plant. A full scale plant with a capacity of 100 000 Mg/y realized in two lines with two pyrolysis drums per line has been built up in Fürth, Germany, and has been commissioned in 1997.

5. Conclusions

The short description of thermal waste treatment processes and the respective associated gas cleaning technologies demonstrates that all different technologies have one common problem: the removal of dust. With the exception of the discussed gasification processes which separate the dust during the quenching this is in all cases done by filtration.

Processes ending up with oxidizing conditions apply in principle multi-functional gas cleaning processes which guarantee the effective abatement of different classes of pollutants. Any acid components are removed by wet or dry scrubbing. Catalytic and non-catalytic abatement strategies for nitrogen oxides are well established. Organics are typically destroyed by (catalytic) oxidation or removed by adsorption on particulate carbon.

Processes under reducing conditions produce gas components which are not that easily categorized and may require specific purification or separation steps for single components like H_2S, HCN or organic products. Hence many processes applied in waste treatment have been modified to end up with an oxidizing environment to minimize the gas cleaning effort. This tendency should be taken into account if thermal treatment processes are considered for the inertization of any waste materials.

6. References

1. Angenend, F.J. & Trondt, L. (1990), Schadstoffinput - Schadstoffoutput, *VGB Kraftwerkstechnik, 70,* 36.
2. Anton, F. (1995), Das Siemens-KWU-Schwelbrennverfahren - Stoffliche und energetische Verwendungen von Restmüll, *VDI-Berichte Nr. 1192,* 349.
3. Berwein, H.-J. & Erlecke, J. (1990), Einsatzmöglichkeiten der Verschwelung als Homogenisierungsstufe in der thermischen Abfallverwertung. In *Müllverbrennung und Umwelt 4 (Thomé-Kozmiensky, K.J., ed.),* Berlin: EF-Verlag, 225.
4. Bilitewski, Härdtle, Marek (1985), Grundlage der Pyrolyse von Rohstoffen. In: *Pyrolyse von Abfällen (Thomé-Kozmiensky, K.J., ed.),* EF-Verlag, Berlin, 1 – 77.
5. Braun, H. (1992), Problemelement Quecksilber - von der Spezifizierung über kontinuierliche Emissionsüberwachung zur Prozeßsteuerung, In *Symposium 25 Jahre LIT, 5 Jahre TAMARA , KfK Report 5000,* 77.
6. Brunner, P.H. & Mönch, H.. (1986), The Flux of Metals Through Municipal Solid Waste Incinerators, *Waste Management & Research, 4,* 105.
7. Buttker, B., Rabe, W. & Müller, J (1993), Kurzbericht über erste Erfahrungen der Verwertung von Reststoffen in der Festbettdruckvergasung in Schwarze Pumpe. VDI Bildungswerk, Mannheim 27./28.1.93.
8. Comfére, W. (1985), Niedertemperaturpyrolyse, BKMI-Verfahren. In *Pyrolyse von Abfällen (Thomé-Kozmiensky, K.J., ed.),* Berlin: EF-Verlag, 142.
9. Dannecker, W. & Hemschemeier, H. (1990), Level of activated-coke technology for flue gas dust collection behind refuse destruction plants looking at the problem frpom the special aspects of dioxin separation. In *Organohalogen Compounds, Vol. 4, Dioxin 90 - EPRI-Seminar (Hutzinger, O. & Fiedler H., ed.),* Ecoinforma Press, Bayreuth, 1990, 267.
10. Fichtel, K. (1987), Bericht über die Müllpyrolyse-Anlage Burgau. In *Müllverbrennung und Umwelt 2 (Thomé-Kozmiensky, K.J., ed.),* Berlin: EF-Verlag, 662 – 678.
11. Häßler, G. (1995), *Thermoselect.* Verlag Karl Görner, Karlsruhe.
12. Hagenmaier, H., Kraft, M., Brunner, H. & Haag, R. (1987), Catalytic effects of Fly Ash from Waste Incineration Facilities on the Formation and Decomposition of Polychlorinated Dibenzo-p-Dioxins and Polychlorinated Dibenzofurancs, *Environmental Science and Technology, 21,* 1080.
13. Hiraoka, M., Takeda, N., Okajima, S., Kasakura, T. & Imoto, Y. (1989), Catalytic Destruction of

PCDDs in Flue Gas, *Chemosphere, 19,* 361.

14. Hunsinger, H., Kreisz, S. & Vogg, H. (1995), Formation of Chlorinated Aromatic Compounds in the Raw Gas of Municipal Solid Waste Incineration Plants, *DIOXIN 95, 15th Symposium on chlorinated Dioxins and Related Compounds, Edmonton, CDN, Short Papers*, 343.
15. International Ash Working Group (IAWG): A.J. Chandler, T.T. Eighmy, J. Hartlén, O. Hjelmar, D.S. Kosson, S.E. Sawell, H.A. van der Sloot, J. Vehlow (1997), *Municipal Solid Waste Incinerator Residues*, Elsevier, Amsterdam-Lausanne-New York-Oxford-Shannon-Tokyo.
16. Jekel, M., Ritz, J. & Vater, C. (1992), Abwässer und deren Behandlung. In *Müllverbrennung und Umwelt 5 (Thomé-Kozmiensky, K.J., ed.),* Berlin: EF-Verlag, 375.
17. Johnke, B. (1994), Schlackeverwertung und -entsorgung unter Beachtung der Vorgaben gesetzlicher und technischer Regelungen, *Beihefte zu Müll und Abfall, 31,* 200.
18. Kaminsky, W., Sinn, H. & Janning, J. (1979), Technische Prototypen für die Altreifen- und Kunststoffpyrolyse, *Chem.-Ing.-Tech., 51*, 419.
19. Korte H.-J. (1994), NEUTREC Verfahren: Trockenverfahren für die Rauchgasreinigung mit Natriumbicarbonat und Recycling der entstehenden Natriumverbindungen. In *VDI-Handbuch, BW 43-59-06.*
20. Kreisz, S., Hunsinger, H. & Vogg, H. (1995), Technical Plastics as PCDD/F Absorbers, *DIOXIN 95, 15th Symposium on chlorinated Dioxins and Related Compounds, Edmonton, CDN, Short Papers* 443.
21. Kürzinger, K. & Stephan, R. (1989), Hydrochloric Acid and Gypsum (Sulphuric Acid) as Utilizable End Products Obtained from the KRC Process for Cleaning Flue Gases from Incinerators. In *Recycling International (Thomé-Kozmiensky, K.J., ed.),* Berlin: EF-Verlag, 1224.
22. Lenz, S (1979), Optimale Abfallnutzung durch neuartige Pyrolyseverfahren, *Umwelt, 4*, 291.
23. Mätzing, H., Namba, H. & Tokunaga, O. (1993), Kinetics of SO_2 removal from flue gas by electron beam technique, *Radiation Physics and Chemistry, 42*, 673.
24. Nielsen, K.K. & Møller, J.T. (1989), Reduction of Chlorinated Dioxins and Furanes in the Flue Gas from Incinerators with Spray Absorbers and Electrostatic Precipitators, *Chemosphere, 19,* 367.
25. Paur, H.R. (1992), Removal of volatile hydrocarbons from industrial offgas. In *Electron Beam and Electrical Discharge Processing (Penetrante, B.M., ed.)* Berlin, Springer Verlag, 77.
26. Paur, H.R. & Mätzing, H. (1993), Electron beam induced purification of dilute off gases from industrial processes and automobile tunnels, *Radiation Physics and Chemistry, 42*, 719.
27. Reeck, G., Schröder, W. & Schetter, G. (1991), Zukunftsorientierte Abfallverbrennung in der MVA Ludwigshafen, *Müll und Abfall, 23,* 661.
28. Sattler, K (1982), *Umweltschutz, Entsorgungstechnik.* Würzburg: Vogel Verlag.
29. Schingnitz, M. & Fritz, P. (1997), Integration von Abfall- und Energiewirtschaft durch Anwendung des NOELL-Konversionsverfahrens, *Chemie Ingenieur Technik, 69*, 820 – 824.
30. South Coast Air Quality Management District (SCAQMD) (1991), *The SCAQMD Screening, Report.*
31. Stahlberg, R. & Feuerriegel, U. (1995), Das THERMOSELECT-Verfahren zur Energie- und Rohstoffgewinnung -Konzept, Verfahren, Kosten -, *VDI-Berichte Nr. 1192,* 319.
32. Tchobanoglons, G., Theisen, H. & Ellassen, R. (1977), *Solid Wastes.* New York: McGraw Hill Inc..
33. Vehlow, J. (1996), Simple, reliable and yet efficient - modern trends in waste incineration. *UTA Technology and Environment, 2/96*, 144.
34. Vogg, H. & Stieglitz, L. (1986), Thermal Behaviour of PCDD/PCDF in Fly Ash from Municipal Incinerators, *Chemosphere, 15,* 1373.
35. Vogg, H., Hunsinger, H. & Stieglitz, L. (1990), Contribution to Solving the Problem of Dioxins Generated during Waste Incineration, *Chemical Engineering Technology,* 13, 221.
36. Vogg, H., Hunsinger, H., Merz, A., Stieglitz, L. & Vehlow, J. (1991), Head-end-Techniken

zur Dioxinminderung. *VDI Berichte Nr. 895,* 193.

37. Vogg, H. (1992), Abfallverbrennung - Fakten und Perspektiven, *BWK/TÜ/UT-Special, Okt. 1992*, M14.
38. Volkman, Y., Vehlow, J. & Vogg, H. (1991), Improvement of Flue Gas Cleaning Concepts in MSWI and Utilization of By-Products. In *Waste Materials in Construction (Goumans, J.J.J., van der Sloot, H.A. & Albers, Th.G., ed.),* Amsterdam: Elsevier Publishers, 145.

INTRODUCTION OF GREEN PLANTS FOR THE CONTROL OF METALS AND ORGANICS IN ENVIRONMENTAL REMEDIATION

TOMAS MACEK, M. MACKOVÁ, J. BURKHARD and KATERINA DEMNEROVÁ

Institute of Organic Chemistry and Biochemistry, Academy of Sciences of the Czech Republic, Flemingovo n. 2, 166 10 Prague
Department of Biochemistry and Microbiology, Faculty of Food and Biochemical Technology, ICT Prague, Technicka 3, 166 28 Prague,
Department of Environmental Protection, Faculty of Chemical Technology, ICT Prague, Technicka 3, 166 28 Prague, Czech Republic

1. Introduction

In this chapter the potential possibilities of phytoremediation will be discussed, with special attention paid to the application of *in vitro* systems for basic research in the role of plants for the remediation of contaminated sites or flows, and in the improvement of their effectiveness. Due to the expense of conventional engineering techniques, reaching over 500 dollars per ton, or a quarter million dollars per acre, it is not surprising that the cleanup of contaminated sites has not been proceeding at a rapid pace [1]. There is an active effort to develop new, more cost effective technologies for remediation of the environment. One of them is known as phytoremediation. Phytoremediation is defined, according to Cunningham and Berti, as the use of green plants to remove, contain or render harmless environmental contaminants [2]. It is important to note, that this includes the use of vegetation for *in situ* treatment of water, sediments and soils. In this process specially selected or engineered plants can be used for extraction of toxic metals from soil or water, including removal of radioactive elements, removal of toxic organic compounds and, if possible, their mineralization.

Various mechanisms are involved in each of one process listed above. Plants remediate organic compounds by direct uptake of contaminants, as summarized by Schnoor *et al.* [3], followed by subsequent transformation, transport and their accumulation in a non-phytotoxic form, which does not necessarily mean non toxic for man. In addition plants support bioremediation by release of exudate and enzymes that stimulate both microbial and biochemical activity in the surrounding soil and mineralization in the rhizosphere. The use of plants as a final water treatment step and for the disposal of sludge resulting from waste water treatment is centuries old [4]. These processes either "decontaminate" the soil, or "stabilize" the pollutant within it, i.e. preventing its migration to a site of actual danger to human health. Specifically, two subsets of phytoremediation are nearing commercialization. First is phytoextraction, in which high biomass metal-accumulating plants and appropriate soil amendments are used to transport and concentrate metals from the soil into the harvestable part of roots and above-ground shoots, which are harvested with conventional agricultural methods.

F.W. Holm (ed.), Effluents from Alternative Demilitarization Technologies, 71–84.

The other is rhizofiltration, in which plant roots grown in water, absorb, concentrate and precipitate toxic metals from polluted effluents. Cunningham and Berti [3] characterized plants as a solar driven pumping and filtering system that has measurable loading and degradation capacity, which can extract and concentrate particular compounds from the environment.

Phytoremediation has been already successfully implemented by more companies and research teams at universities or in the army. Trees of the *Salicaceae* family (willow and poplar) have been planted at several locations because of their flood tolerance and fast growth. Schnoor *et al.* [4] used hybrid poplars Imperial Carolina for increasing soil suction and decreasing downward migration of pollutants. This system was used to control agricultural runoff along a small creek in Iowa, and as final cap on a landfill in Oregon. The University of Washington has developed multiple hybrid poplar clones, some with growth rates of 10-15 feet per year. These trees were found to have remarkable ability to take up and degrade certain halogenated organic solvents. In laboratory conditions the poplar tissues were used to examine the metabolism of trichloroethylene (TCE). Also the US Air Force is trying to clean TCE from groundwater using poplar trees and the US Army to clean 2,4,6,-trinitrotoluene (TNT) and hexahydro-1,3,5,-trinitro-1,3,5,-triazine (RDX) from contaminated wetlands, using a variety of plants. In pilot sites at numerous industrial facilities for petroleum hydrocarbons and wood preservatives, phytoremediation using various grasses (and their associated bacterial populations) has also had limited success [5]. Care should be taken, however, to choose the proper plant species. For example at the Naval Air Station Joint Reserve Base Fort Worth (formerly Carswell Air Force Base), where eastern cottonwood trees were planted to clean up trichloroethylene from a shallow, thin aerobic aquifer, the organism was chosen, because: "they are (eastern cottonwoods) indigenous to the site, they are members of the same genus as poplar trees and grow at similarly rapid rates"! The trees were planted to intercept and reverse the flow gradient and determine if cottonwoods can metabolize TCE and its daughter compounds under field conditions [6]. The systems is working well, decreased effectively the TCE content, but from the published information the above explanation of reasons leading to the choice of eastern cottonwood seems to be quite unsatisfactory, missing support of research results. In the above case the good results proved the choice was right, but in general the approach cannot be considered as optimal. This is often the problem of agencies funding remediation experiments, time pressure leads sometimes to support of projects missing the time-consuming attempts to explain experimentally the basis of molecular mechanisms involved in the desired activity.

1.1 ADVANTAGES OF PHYTOREMEDIATION

According to Salt *et al.* [7], and [1, 3, 8,], the main advantages of phytoremediation in comparison with classical remediation methods can be easily summarized:

- it is far less disruptive to the environment.
- there is no need for disposal sites.
- it has a high probability of public acceptance.

- avoids excavation and heavy traffic.

Considering these factors, and the much lower cost expected for phytoremediation, it appear that it will be used in much larger scale clean-up operations than possible by other methods. The process is relatively inexpensive, because it uses the same equipment and supplies used in agriculture.

1.2 DISADVANTAGES OF PHYTOREMEDIATION

Like other methods of environmental remediation, also phytoremediation has its disadvantages. These vary with the application and type of contamination for which the method will be used [1, 3, 7, 8], the main concerns come from the following problems:

- formation of vegetation may be limited by extremes of environmental toxicity.
- contaminants collected in leaves can be released again to the environment during litter fall.
- contaminants can be accumulated in fuel woods.
- some contaminants may have increased solubility resulting in greater environmental damage and/or pollutant migration.
- it may take longer than other technologies.

Many of these problems can be addressed by the selection of proper type of plants. Phytoremediation has a number of inherent technical limitations. The contaminant must be within, or drawn toward, the root zones of plants (usually the top three to six feet of soil). This implies water, depth, nutrient, atmospheric, physical and chemical limitations. In addition, the site must be large enough to make farming techniques appropriate. It must not present an eminent danger to human health or further environmental harm [4]. There may be also a considerable delay in time needed for obtaining satisfactory clean-up results between phytoremediation and "dig and dump" techniques. Although plants have upwards of 100 million miles of roots per acre, their root system may not extend deeply enough to eliminate all of the contamination [5].

1.3 FUTURE PROSPECTS

The growing knowledge of the factors important to phytoremediation can provide a basis for genetic modification of plants for improved performance. Breeders have been modifying agronomically important plant traits for years. However, in these instances yield and aesthetics were usually the criteria for selection of a plant trait. As Cunningham and Berti [2] state, "phytoremediation requires a new paradigm in which plants are valued based on what they adsorb, sequester, destroy and tolerate". Traditional breeding as well as molecular biology can specifically target all of these traits. We believe that molecular biology will allow formation of plants tailored specifically for needs of some particular tasks, e.g. as described by Macek *et al.* [9].

These changes will include transforming of the plants to add specific proteins or peptides for binding and transporting xenobiotics, increasing the quantity and activity of plant biodegrading enzymes (peroxidases, laccases, oxygenases), including those that are exported into the rhizosphere and surrounding soil in order to improve the performance of soil bacteria. Introducing foreign genes might include those responsible for the synthesis of low molecular weight organic molecules to be excreted in exudates, like some phenolics, flavonoids or coumarins as described by Fletcher *et al.* [10, 11], that induce rhizospheric bacteria to degrade anthropogenic toxins. Plant-fungal interactions would also seem ripe for exploitation in this area, particularly mycorrhizal associations [12]. Alternatively, transgenic plants can be transformed to harbor microbial genes for biodegradation. This is already a routine practice in the engineering of many herbicide-resistant plants and their field testing, product development, and registration are well advanced. The concept could be extended to address additional xenobiotics. The case of biodegrading microbial strains is notoriously unreliable in ability to compete with native microorganisms in the natural environment [12]. In addition, such release faces general public opposition. In contrast to herbicide resistance, the degradation of xenobiotics is rarely performed by one single enzyme. For the degradation of organic xenobiotics concerted action of many enzymes is needed, and introduction of all of the necessary genes into a plant is a formidable task.

Exaggerated claims of success in the early years of bacterial remediation led to inflated expectations for the use of phytoremediation. Therefore, it is incumbent upon investigators studying phytoremediation to conduct thorough basic research before attempting field implementation [5].

2. *In vitro* cultured plant cells in phytoremediation studies

Most of the experiments used for establishing the phytoremediation techniques were done with normal soil-grown plants or hydroponically. Nevertheless, as more and more efforts is drawn toward the basic research to understand and improve the performance of plants in remediation technologies, the number of results obtained with the help of *in vitro* plant cell and tissue cultures is rapidly increasing. The concept is not new, *in vitro* cultivated plant cells have been used in studies of herbicide resistance and metabolism for many years, also other organic xenobiotics were studied, e.g. metabolism of pentachlorophenol by wheat and soybean cell suspension cultures yields glucosides and non-extractable residues, as described by Langebartels and Harms [13]. The metabolism of PCBs by plant cells was addressed already 10 years ago by Groeger and Fletcher [14]. Metabolism of many other organic compounds has been addressed with the help of plant cell and tissue cultures. Plant cell cultures serve also as a tool in studies of enzyme systems involved in the metabolism of xenobiotics [15, 16, 17]. It was proved that such cultures are able to metabolize the compounds by common metabolic pathways and can be used as a useful system with some advantages in comparison with intact plants [18]. The material can be grown under standard laboratory conditions, the growth is independent of the weather and often *in vitro* cultures grow more rapidly. In studies of heavy metal uptake the *in vitro* cultured plant cells and tissues play also an important role, as shown e.g. by Macek *et al.* [9, 19, 20] in studies of cadmium uptake. The exploitation of transformed hairy root cultures is especially rewarding, as discussed by Macková *et al.* [21, 22], Hughes *et al.* [23] or Betts [24], because these *Agrobacterium rhizogenes* transformed plant roots exhibit all features of normal plant roots and grow rapidly under defined aseptic conditions *in vitro*, thus allowing to distinguish between the plant metabolism itself and the effect of the complex interaction between plants and microbial communities in the rhizosphere. *In vitro* techniques using plant cell, tissue and organ cultures have been most commonly applied for studying the fate of xenobiotics in plants.

3. Contamination by heavy metals

Certain bacterial, fungal, algal and plant systems are capable of concentrating some toxic inorganics. Extraction of inorganic contaminants from soil by these organisms is theoretically possible, however, if the microorganisms are considered, no cost effective way currently exists to remove them from the soil after they sequester inorganic ions. Harvesting plants, on the other hand, is a familiar agricultural technology [4]. The situation looks better if contaminated water is considered. In such case a wider range of organisms seems to be promising, from water plants to microalgae, from root filters to immobilized bacteria. Waste products such as sludge, ash, waste water and mine drainage water and urban refuse city damps, have high nutrient values but often also a high heavy metal content. In most applications of plants in phytoremediation the tolerance to heavy metals has to be considered. This aspect is in depth discussed by Shaw in [25].

Different plant species are tested for their usefulness in phytoremediation. For example, willow (*Salyx*) has a high capacity to accumulate heavy metals compared to other species. Landberg and Greger [26] screened 103 different clones of *Salyx* of various origins. Some clones were tolerant to both Zn and Cd, some only to one of them. Transportation of metal ions to the shoots varied between 1 and 72% of the total metal uptake. Tolerant clones were found either having high accumulation in roots, or having a low net metal ion uptake. The latter were probably metal ion tolerant because they excluded or exuded the metals. According to the authors the most important parameter is not the plants tolerance to heavy metals, but the effectiveness of a clone in the accumulation or exclusion of heavy metals. In addition to accumulation capacity, biomass production must be considered in order to determine the total metal uptake.

3.1 HYPERACCUMULATORS

Finding or developing plants that acquire high levels of metal contaminants in harvestable tissue was thought impossible until the (re)discovery of a small group of remarkable plants called hyperaccumulators, described by Brooks *et al.* [27]. These plants are relatively uncommon, but widespread in the plant kingdom. Hyperaccumulating species can be found among the heavy metal tolerant plants, which were used for identification of localities with high ore content. Early attention to the special morphology of metallophytic plants gave already the Czech geobotanist Novák [28] in 1928, or Italian Pichi-Sermoli [29]. A complete overview is given in [25].

TABLE I. Metal concentration (on a dry weight basis) in some known hyperaccumulators, as summarized by Cunningham and Ow [4].

Metal	Species	Concentration in shoots (DW)
Mn	*Macadamia neurophylla*	51,800 mg kg^{-1}
Zn	*Thlaspi caerulescens*	51,600 mg kg^{-1}
Ni	*Psychotria douarrei*	47,500 mg kg^{-1}
Cu	*Ipomoea alpina*	12,300 mg kg^{-1}
Co	*Haumaniastrum robertii*	10,200 mg kg^{-1}
Pb	*Thlaspi rotundifolium*	8,200 mg kg^{-1}
Cd	*Thlaspi caerulescens*	1,800 mg kg^{-1}

3.2 PHYTOEXTRACTION

The processes involved in phytoextraction are nearly the same for either contaminated soil or water. The contaminant must be in a biologically accessible form, and root absorption must be possible and must occur. Translocation of the contaminant from root to shoot makes tissue harvesting easier and lessens exposure to the contaminant. The actual rate of removal of a contaminant from a contaminated environment is dependent on the biomass gathered during harvesting, the number of harvests per year, and the metal concentration in the harvested portion of the plant. Decontamination of a site requires a plant with both a high biomass yield and high metal accumulation of 1 to 3 % metal by dry weight [4]. Although some of the properties desired for phytoremediation are present in currently recognized hyperaccumulators, many have low biomass and restricted element selectivity in hyperaccumulation. In addition there is little or no information available about the agronomics, genetics, breeding potential, and disease spectrum of most of these plants.

After harvesting, a processing step is necessary to recover most of the metal contaminants. Alternatively, the harvested biomass could be reduced in volume and/or weight by thermal, microbial, physical or chemical means (composting, anaerobic digestion, low temperature ashing, leaching). This step would further decrease the costs of handling and processing, as discussed by Salt *et al.* [7]. After stabilization potential subsequent landfilling can be considered. With some metals, after leaching or smelting in kiln (e.g. Ni, Zn, and Cu), the value of the reclaimed metal may provide an additional incentive for remediation. This step reduces generation of hazardous waste and generates recycling revenues.

3.3 THE ROLE OF AMMENDMENT

Huang and coworkers [30] investigated the potential of adding chelates to Pb-contaminated soil to increase Pb accumulation in plants. Calculations of soil Pb mass balance suggest that this technology will be economically feasible only if systems can be developed to employ high biomass plants that can accumulate greater than 1% Pb in their shoots. The addition of chelates to soil (total soil Pb 2500 mg kg^{-1}) increased shoot Pb concentrations of corn and pea from less than 500 mg kg^{-1} to more than 10 000 mg kg^{-1}. The surge of Pb accumulation in these plants was associated with the surge of Pb level in the soil solution due to the addition of chelates. The results indicate that chelates enhanced Pb desorption from soil to soil solution, facilitated Pb transport into the xylem, and increased Pb translocation from roots to shoots. The authors suggest that with careful management, chelate assisted phytoextraction may provide a cost-effective soil decontamination strategy. On the other hand, tailoring the project to local conditions, can according to Brown *et al.* [31] help reducing the toxic effect, e.g. mixing high iron and high lime sewage sludge compost into high zinc soil reduces zinc toxicity to grasses, thus allowing growing lawns in Palmerton, Pennsylvania area. Without these additives, zinc contamination causes serious growing problems for vegetation in area, contaminated probably by a zinc smelter operating between 1890 and 1980-before modern environmental regulations existed. Tall fescue varieties grew very nicely on soil treated with sludge compost. Of the plant species tested at Pig´s Eye Landfill in St. Paul, Minnesota, alpine pennicress was the best at taking in cadmium, zinc, and lead [32]. Pennicress has proven especially good at removing zinc and cadmium, accumulating up to 30,000 ppm of zinc in its leaves without yield reduction. Zinc is taken up at the rate of 125 kilograms per hectare per year, if fertilized and managed carefully.

4. Contamination by organic compounds

Plants can accelerate bioremediation in surface soils by their ability to stimulate soil microorganisms through the release of nutrients from and the transport of oxygen to their roots [1, 3]. The zone of soil closely associated with the plant root, the rhizosphere, has much higher numbers of metabolically active microorganisms than unplanted soil. The interaction between plants and microbial communities in the rhizosphere is complex and has evolved to the mutual benefit of both organisms. This symbiotic relationship is responsible for the accelerated degradation of soil contaminants in the presence of plants.

In addition to this rhizosphere effect, plants themselves are able to passively take up a wide range of organic wastes from soil or water through their roots. In this paper we shall not discuss the uptake of organic contaminants from air, but it has to be stated, that plants have a great capacity to bind organic compounds from air.

Classes of organic compounds that are more rapidly degraded in rhizosphere than in unplanted soil include polycyclic aromatic hydrocarbons, total petroleum hydrocarbons,
chlorinated pesticides (PCP, 2,4-D), other chlorinated compounds (PCBs, TCE), explosives (TNT, RDX, DNT), organophosphate insecticides (diazinon, parathion) and surfactants (detergents).

Kansas State University researchers [33] proved in field test substantial decrease of total petroleum hydrocarbon content using Bermuda grass, rye grass, white clover and tall fescue in plots at US Navy´s largest fuel facility in the United States, at Craney Island Fuel Terminal. The removal of trichloroethylene was used as an example already at the beginning of this chapter, many laboratories deal with removal of explosives. There are numerous Department of Defense sites across the USA with explosives contaminated groundwater. The U.S. Army environmental Center (USAEC) is developing technologies to effectively clean up groundwater contaminated with residues of explosives like TNT, RDX, HMX, and DNT. Current groundwater cleanup technologies, such as granular activated carbon and advanced oxidation, have proven to be labor intensive and costly. One potential treatment alternative is phytoremediation using constructed wetlands [34]. The EPA National Exposure Research Laboratory identified in bench scale testing a plant nitroreductase, which in cooperation with other plant enzymes is able to degrade TNT, RDX and HMX. An artificial wetland for demonstration of the feasibility of using selected plants to cleanup explosives-contaminated groundwater was constructed at Milan Army Ammunition Plant, TN, experiments begun in 1995. A pilot-scale plant lagoon system was constructed and operated by Georgia Institute of Technology. Results indicate that the TNT removal in plant cells matched laboratory batch study predictions. Removal percentages relative to TNT loading for all species ranged from a low 85.4% to a high of 99.7%.

4.1 METABOLISM OF POLYCHLORINATED BIPHENYLS BY PLANT CELLS

It was shown that plants are able to degrade PCBs, but the data for such models are scarce comparing to the vast amount of papers on bacterial or fungal pathways of PCB degradation. In the paper of Lee and Fletcher [15] the authors suggested that rather than peroxidases, cytochrome P450 is involved in the PCB degradation pathway. Nevertheless, we found during incubation of cultures in the presence of PCB a positive correlation between peroxidase content and PCB disappearance [16, 17]. For general experiments with metabolisation of PCBs usually standard mixtures of congeners are used, but to investigate the fate of PCBs in the plant cell, pure individual congeners have to be tested. Butler *et al.* [44] demonstrated, that plant cells are capable of hydroxylating and glycolsylating a monochlorinated biphenyl in a manner similar to what has been reported for animals. Wilken *et al.* [45] followed the metabolism of 10 different congeners of PCBs, with cultures of 12 plant species. The metabolism of defined congeners was strongly dependent on the plant species. In particular, two *Fabaceae* species exhibited high capacities for metabolizing different congeners. The substitution and molecular configuration play an important role in PCB metabolism. The hydroxylation pattern was established in some cases. Groeger and Fletcher [14] showed, that metabolism of PCB congeners by axenic plant cultures was inversely related to the chlorine content of the congener. Further studies on improvement of PCB degradation by plants, e.g. cooperation with mycorrhizal fungi, support of the growth of PCB-degrading bacteria by compounds from photosynthetic plants described by Fletcher and coworkers are described in part 1.3 of this chapter [10, 11, 12].

In the Czech Republic contamination by PCBs is a serious problem. For this reason different approaches to their removal from environment are under investigation, including the use of plants, as summarized by Demnerova et al. [43] or Kas *et al.* [41]. In our experiments we studied the ability of the cultures of various species cultivated *in vitro* to degrade Delor 103 - a mixture of PCBs formerly commercially produced in the former Czechoslovakia. The PCBs mixture contains about 60 congeners with a different degree of chlorination by the average number of 3 chlorines per biphenyl molecule. The analytical procedure was optimalised by Burkhard *et al.* [42]. We tested first cultures of *Solanum aviculare* [40], than further cultures of different plant species cultivated *in vitro* [21] and evaluated their PCB degradative abilities in relation to their origin and morphology [17]. Altogether already over 40 cultures were screened. The effect of PCBs on the changes of the level of peroxidase activity and the pattern of peroxidase isoenzymes was also followed [16, 17]. Better yields of PCB metabolisation were obtained with cultures more differentiated morphologically, and cultures transformed by *Agrobacterium tumefaciens* or *A. rhizogenes* were also more active than non transformed ones of the same species, established Mackova *et al.* [17]. Very effective proved to be hairy root cultures, tested by Mackova *et al.* [21], obtained after transformation by *A. rhizogenes*. From selected hairy root cultures of different species the performance of most active one, *Solanum nigrum* clone SNC-9O was studied in more detail by Mackova *et al.* [22]. The hairy root cultures are a very good model for phytoremediation studies, resembling all main features of a normal plant root, but having very good growth properties, thus allowing faster *in vitro* experiments on a fully differentiated experimental material.

5. Enhancement of phytoremediation by breeding or genetic engineering

There are many attempts to breed willow, poplar and other plants mentioned in this article with properties useful for phytoremediation purposes. Classical breeding and selection of strains bearing the necessary enzyme systems, resistance against toxic compounds in the environment and satisfactory biomass production is a very difficult task. This can take many years to obtain a plant with the required properties. On the other hand, there are very promising results obtained by the means of genetic engineering. This is especially remarkable in the field of herbicide resistance, but also genes responsible for other important traits are being introduced into plant species which might prove useful in phytoremediation.

5.1 ATTEMPTS TO INCREASE METAL BINDING IN PLANTS

There are more different mechanisms involved in exclusion or hyperaccumulation of heavy metals by plants. These include the uptake itself, including the effect of exudates, transport, storage in different forms and different organs, and resistance. One way is to find the gene or genes responsible for hyperaccumulation and insert them into a high yielding plant. The other option is to use traditional breeding to produce e.g. alpine pennycress plants that grow faster and taller.

An interesting approach based on inserting an altered mercuric ion reductase gene (*merA*) into *Arabidopsis thaliana*, was successfully tested by Meagher and coworkers at the University of Georgia. This way a mercury-resistant transgenic plant was produced, that volatilizes mercury into the atmosphere. Transgenic *merA* plants are soon to be tested in soil [5].

Plants have developed their own systems for binding of heavy metals based largely on the synthesis of phytochelatins, described by Grill *et al.* [35]. The presence of heavy metals, e.g. Cd, is an important stress factor, influencing the enzymatic activities of the plant cell [16]. Metallothioneins are a group of proteins with high affinity to heavy metals, present in many organisms. In plants mostly metallothioneins of group III were described, but plants are able also to produce group I and II metallothioneins, as reviewed in [36]. With the aim to improve the metal binding ability, some foreign genes coding for metallothioneins were already introduced by genetic engineering into plants. The introduction of the yeast metallothionein gene CUP1 (yeast copper-binding protein) might improve the ability of a plant to tolerate or accumulate heavy metal ions. Its introduction and expression into tobacco plants has been recently described. The strategy to create plants combining high metal accumulating capacity and suitable agrotechnical characteristics has been approached at the Biotechnology Center of Utah State University in Logan. Specifically, the yeast metallothionein gene, CUP1, was genetically introduced by Truksa *et al.* [37] into plants in an attempt to improve their ability to tolerate or accumulate heavy metal ions. The gene coding for the CUP1 protein was placed under the control of CaMV 35S promoter and tobacco was transformed with this construct using *Agrobacterium tumefaciens*. The selected transformants of F_1 generation are characterized by higher tolerance towards toxic metals in the early stages of development and the preliminary results of metal uptake experiments showed also increased capacity of metal accumulation.

The introduction of an additional metal binding domain to the implemented protein should further enhance the metal binding capacity, as tested by Macek *et al.* [38]. For this reason a gene coding a peptide of six histidine moieties has been isolated from a commercial bacterial expression plasmid, pTrc-HisA. This polyHis domain was originally designated for the purpose of easy and effective isolation of engineered proteins by selective binding on a metal containing carrier using immobilized metal affinity chromatography (IMAC). The gene was cloned in *E. coli* using plasmid pBI121 bearing CaMV 35S promoter and β-glucuronidase (GUS) reporter gene. In the next step the GUS gene from the new construct pBI-HisGUS was replaced by the CUP1 gene. The prepared plasmid was electroporated into *Agrobacterium* strain LBA4404 and used for transformation of tobacco (*Nicotiana tabacum* Wisconsin 38). The best yield of regenerated transformed plantlets was obtained by Macek *et al.* [9] on LS medium supplemented by naphtaleneacetic acid and benzylaminopurine]. To evaluate the effect of different constructs on the interaction of bacterial cells with heavy metals, *E. coli* cultures bearing expression vectors containing GUS, HisGUS and HisCUP were used to follow the expression of recombinant proteins, their effect on culture growth and metal uptake. These parameters were compared using media with high concentrations of Ni, Cu, Cd and Zn.

<u>HisCUP protein expression in bacteria</u>

The expression of recombinant proteins, and the growth and heavy metal accumulation in bacteria *E. coli* TOP10 bearing expression vectors containing GUS, HisGUS and HisCUP was studied by cultivation at 100 rpm, 37°C, in 250 ml flasks with 100 ml LB medium. Bacterial growth was followed by optical density at 600 nm. Overnight cultures were diluted to an OD of 0.100 using either fresh LB medium or LB medium containing a non-toxic (0.25 mM) concentration of cadmium. Isopropyl-β-D-thiogalactoside (IPTG), if added, was supplied to the bacterial culture at OD = 0.400, to a final concentration of 1 mM, to induce the expression of the recombinant fusion protein. The samples were then grown until absorbance reached 0.6-0.7 and cadmium concentration was increased to a toxic level (2 mM). Aliquots of the bacterial suspension were taken at regular intervals, between 0.5 to 30 hours. Protein expression was detected immunochemically [9].

Transformation of the plants was confirmed by plasmid-located kanamycine resistance of transformed plants and by electrophoretic identification of the polyHisCUP1 DNA fragment after PCR amplification of cDNA. This was obtained by reverse transcription of mRNA isolated from the leaves of transgenic plants.

Expression of the recombinant fusion protein and cadmium accumulation was tested in *E. coli* TOP10 bearing prepared constructs. Over 160 full scale growth curves were measured for bacteria containing plasmids with inserts HisCUP, HisGUS, GUS and controls [9].

TABLE II: Cadmium accumulation (mg.kg^{-1} dry weight) following the addition of + 0.25 mM Cd + IPTG + 2 mM Cd (average from 3 independent experiments)

sample	TOP10	pSE420	GUS	HisGUS	HisCUP
cadmium accumulation mg.kg^{-1} DW	5500	9080	10300	13570	12760

TABLE III: Cadmium accumulation (mg.kg^{-1} dry weight) by *E. coli* TOP10 containing HisCUP insert

addition of cadmium	+2mMCd	+ IPTG + 2mMCd	+ 0.25 mM Cd +IPTG +2mM Cd
accumulation mg.kg^{-1} DW	5700	8090	11 130

Accumulation of heavy metals increased significantly in bacterial cells bearing the transgene coding for the polyhistidine cluster, while an increase in the resistance (ability to grow) towards metals in media was not remarkable for cultures bearing transgenes with His or HisCUP. The accumulation was particularly high in the clone bearing the HisCUP insert.

We have currently available tobacco plants bearing CUP, GUS, HisGUS and HisCUP genes under a constitutive promoter, thus enabling us to follow their heavy metal tolerance and uptake changes as function of the transgene present [9, 37, 38]. The transformation may be applied also to other species, and we are optimistic that this may lead to the development of plants suitable for bioremediation.

6. Conclusions

Hybrid technologies such as phyto-vapor extraction are being actively investigated for technical and economic feasibility. The use of plant roots as "biocurtains" or "biofilters" for the passive remediation of shallow groundwater is also an active area of research [4]. These hybrid technologies will undoubtedly provide both research opportunities and potential field applications in the short term. In the longer term, use of such hybrid technologies will probably have significantly greater costs than is likely achievable by phytoremediation utilizing advanced plant selection and genetic engineering techniques.

According to the American Society of Plant Physiologists (ASPP), the phytoremediation cost in the case of contaminated soil is within 15 to 30 dollars per one ton, while conventional methods (removal and incineration) cost between 500 to 1000 USD per one ton of soil [39]. In contrast to stripping the contaminants from the soil using physical, chemical or thermal processes, transfer of soil to hazardous waste landfills etc., such an *in situ* approach like phytoremediation has additional advantages. The establishment of vegetation on a site also reduces soil erosion by wind and water, which helps to prevent the spread of contaminants and reduces exposure of humans and animals.

There have been about two dozen field tests of phytoremediation to date, with many new appearing, yet, in many ways it is still in its infancy. A comforting thought for plant biologists is that much of the research effort will center on a deeper understanding of basic plant processes [4]. New innovative methods are needed to solve some of our worst pollution problems. The concept of manipulating plant genes that regulate toxic metal uptake is cutting-edge research. The likelihood of public acceptance and the fact that permits of field testing of genetically engineered plants now vastly outstrip permits for genetically altered microbes represent a good reason for moderate optimism.

7. Acknowledgements

Our results described in this paper were supported by the grant 203/96/0499 of the Grant Agency of the Czech Republic, by internal grant of Utah State University Biotechnology Center, Logan, Utah, USA and by grant COST 66-20.

8. References

1. Cunningham S.D., Berti W.R. and Huang J.W. (1995) Phytoremediation of contaminated soils. *TIBTECH* **13**, 393-397
2. Cunningham S.D. and Berti W.R. (1993) Remediation of contaminated soils with green plants: an overview. *In Vitro Cell Dev.* Biol. **29P**, 207-212
3. Schnoor J.L., Licht L.A., McCutcheon S.C., Wolfe N.L., and Carreira L.H. (1995) Phytoremediation of organic and nutrient contaminants. *Environmental Science and Technol.* **29**, A318-323
4. Cunningham S.D. and Ow D.W. (1996) Promises and prospects of phytoremediation. *Plant Physiol.* **110**, 715-719
5. Boyajian G.E. and Carreira L.H. (1997) Phytoremediation: A clean transition from laboratory to marketplace? *Nature Biotechnology*, **15**, 2, 127-128
6. Betts K.S. (1997) Phytoremediation project taking up TCE. *Environmental Science and Technol.* **31**, 8, 347A
7. Salt D.E., Blaylock M., Kumar N. P.B.A., Dushenkov V., Ensley B.D., Chet I. and Raskin I. (1995) Phytoremediation: A novel strategy for the removal of toxic metals from the environment using plants. *Biotechnology*, **13**, 5, 468-474

8. Anderson T.A., Guthrie E.A., Walton B.T. (1993) Bioremediation. *Environmental Science and Technol.* **27**, 13, 2630-2636
9. Macek T., Mackova M., Kotrba P., Truksa M., Singh-Cundy A., Scouten W.H. and Yancey N. (1997) Attempts to prepare transgenic tobacco with higher capacity to accumulate heavy metals containing yeast metallothionein combined with a polyhistidine, in H. Verachtert and W. Verstraete (eds.), *Environmental Biotechnology*, Proc. Int. Symp., Oostende, April 1997, Technological Institute Gent, pp. 263-266
10. Fletcher J.S., Donnelly P.K. and Hegde R.S. (1995) Biostimulation of PCB-degrading bacteria by compounds released from plant roots, in R.E. Hinchee, D.B. Anderson and R.E. Hoeppel (eds.), Bioremediation of recalcitrant organics, Battelle Press, Columbus, pp. 131-136
11. Hegde R.S. and Fletcher J.S. (1996) Influence of plant growth stage and season on the release of root phenolics by mulberry as related to development of phytoremediation technology. *Chemosphere*, **32**, 12, 2471-2479
12. Donnelly P.K. and Fletcher J.S. (1994) Potential use of mycorrhizal fungi as bioremediation agents, in T.A. Anderson, (ed.), *Bioremediation through rhizosphere technology*, ACS Symposium Series No. 563, American Chemical Society, pp. 93-99
13. Langebartels C. and Harms H. (1984) Metabolism of pentachlorophenol in cell suspension cultures of soybean and wheat: Pentachlorophenol glucoside formation. *Z. Pflanzenphysiol.* **113**, 3, 201-211
14. Groeger A.W. and Fletcher J.S. (1988) The influence of increasing chlorine content on the accumulation and metabolism of polychlorinated biphenyls by Paul´s Scarlet Rose cells. *Plant Cell Reports*, **7**, 329-332
15. Lee I. and Fletcher J.S. (1992) Involvement of mixed function oxidase systems in PCB metabolism by plant cells. *Plant Cell Reports*, **11**, 97-100
16. Macek T., Macková M., O enášková J., Demnerová K., Pazlarová J., K en V. (1996) Peroxidase isoenzyme patern and total activity changes in plant cells cultivated *in vitro* under abiotic stress, in C. Obinger, U. Burner, R. Ebermann, C. Penel, H. Greppin (eds.), *Plant Peroxidases: Biochemistry and Physiology*, University of Geneva, pp. 380-385
17. Macková M., Macek T., Burkhard J., O enášková J., Demnerová K., Pazlarová J. (1997) Biodegradation of polychlorinated biphenyls by plant cells. *International Biodeterioration and Biodegradation*, **39**, 4, 317-325
18. Harms H. and Kottutz E. (1990). In: *Progress in Plant Cellular and Molecular Biology*, H.J.J. Nijkamp, L.H.W. van der Plas and J. van Aartrijk, (eds.), Kluwer Academic Publishers, Dordrecht Boston, pp. 650-655
19. Macek T., Kotrba P., Š chová M., Skácel F, Demnerová K. and Ruml T. (1994) Accumulation of cadmium by hairy root cultures of *Solanum nigrum. Biotechnol. Letters*, **16**, 6, 621-624
20. Macek T., Kotrba P., Ruml T., Skácel F. and Macková M. (1994) Accumulation of cadmium by hairy root cultures, in P.M. Doran (ed.), *Hairy Roots: Culture and Application*, Harwood Academic Publishers, pp. 133-138
21. Macková M., Macek T., O enášková J., Burkhard J., Demnerová K., Pazlarová J. (1996) Selection of the potential plant degraders of PCB. *Chemické listy*, **90**, 9, 712-713
22. Macková M., Macek T., Ku erová P., Burkhard J., Pazlarová J. and Demnerová K. (1997) Degradation of polychlorinated biphenyls by hairy root culture of *Solanum nigrum. Biotechnol. Letters*, **19**, 8, 787-790
23. Hughes J.B., Shanks J., Vanderford M., Lauritzen J. and Bhadra R. (1997) Transformation of TNT by aquatic plants and plant tissue cultures. *Environmental Science and Technol.* **31**, 266-271
24. Betts K.S. (1998) Getting to the root of phytoremediation. *Environmental Science and Technol.* **32**, 1, A18
25. Shaw A.J. (1990) *Heavy metal tolerance in plants: Evolutionary aspects.* CRC Press, Boca Raton.
26. Landberg T. and Greger M. (1994) Can heavy metal tolerant clones of *Salyx* be used as vegetation filters on heavy metal contaminated land? In: P. Aronsson and K. Perttu, (eds.), *Willow vegetation filters for municipal wastewaters and sludges*, S.U.A.S., Uppsala, 133-144
27. Brooks R.R., Lee J., Reeves R.D., Jaffre T. (1977) Detection of nickeliferous rocks by analysis of herbarium specimens of indicator plants. *J. Geochem. Expl.* **7**, 49-77
28. Novák F.A. (1928) Quelque remarques relatives au probleme de la vegetation sur les terrains serpentiniques. *Preslia*, **6**, 42-48
29. Pichi-Sermolli R. (1948) Flora e vegetazione delle serpentine e delle altre ofioliti dell alta valle del Trevere (Toscana). *Webbia*, **6**, 1-9
30. Huang J.W., Chen J.J., Berti W.R., and Cunningham S.D. (1997) Phytoremediation of lead-contaminated soils: Role of synthetic chelates in lead phytoextraction, *Environmental Science and*

Technol. **31**, 800-805

31. Brown S.L., Chaney R.L., Angle J. S. and Baker A.J.M. (1995) Zinc and cadmium uptake by hyperaccumulator *Thlaspi caerulescens* and metal tolerant *Silene vulgaris* grown on sludge-ammended soils. *Environmental Science and Technol.* **29**, 1581-1585
32. Commis D. (1995) Metal-scavenging plants to cleanse the soil. *Agricultural Research, USDA-ARS*, **11**, 4-9
33. Betts K.S. (1997) TPH soil cleanup aided by ground cover. *Environmental Science and Technol.* **31**, 5, 214A
34. Betts K.S. (1997) Native aquatic plants remove explosives. *Environmental Science and Technol.* **31**, 7, 304A
35. Grill E., Loffler S., Winnacker E-L. and Zenk M.H. (1989) Phytochelatins, the heavy-metal-binding peptides of plants, are synthetised from glutathione by a specific γ-glutamylcysteine dipeptidyl transpeptidase (phytochelatin synthase). *Proc. Nat. Acad. Sci USA*, **84**, 6838-6846
36. Kotrba P., Ruml T. and Macek T. (1994) Cadmium binding by microbial and plant cells. *Chem. Listy*, **88**, 642-649
37. Truksa M., Singh-Cundy A., Macek T., Kotrba P., Macková M., Yancey N., Scouten W.H. (1996) Transgenic plants expressing metal binding proteins in phytoremediation. *Chem. Listy*, **90**, 9, 707
38. Macek T., Macková M., Truksa M., Singh-Cundy A., Kotrba P., Yancey N., Scouten W.H. (1996) Preparation of transgenic tobacco with a yeast metallothionein combined with a polyhistidine tail. *Chem. Listy*, **90**, 9, 690
39. Jones R.L. (1994) ASPP recomends hazardous waste remediation technologies to DOE. *ASPP Newsletter*, **21**, 6, 12-13
40. Macek T. (1989). Poroporo, *Solanum aviculare*, *S. laciniatum*: *in vitro* culture and the production of solasodine, in *Biotechnology in Agriculture and Forestry*, **7**, Y.P.S. Bajaj (ed.), Springer Verlag Berlin Heidelberg New York Tokyo, 443-467
41. Káš J., Burkhard J., Demnerová K., Koš ál J., Macek T., Macková M. and Pazlarová J. (1997) Perspectives in biodegradation of alkanes and PCBs. *Pure ans Appl. Chem.* **69**, 11, 2357-2369
42. Burkhard J., Macková M., Macek T., Ku erová P. and Demnerová K. (1997) Analytical procedure for the estimation of PCB transformation by plant tissue cultures. *Anal. Commun.* **34**, 10, 287-290
43. Demnerová K., Burkhard J., Koš ál J., Macková M., Pazlarová J., Kuncová G., Macek T. and Kaštánek F. (1997) Biodegradation of alkanes and PCBs: Experience in the Czech Republic, in F.W. Holm (ed.), *Mobile Alternative Demilitarization Technologies*, NATO ASI Series 1, Vol. **12**, 53-70
44. Butler J.M., Groeger A.W. and Fletcher J.S. (1992) Characterization of monochlorinated biphenyl products formed by Pauls Scarlet Rose cells. *Bull. Environ. Contam. Toxicol.* **49**, 821-826
45. Wilken A., Bock C., Bokern M. and Harms H. (1995) Metabolism of different PCB congeners in plant cell cultures. *Environ. Toxicol. and Chem.*, **14**, 12, 2017-2022

REMOVAL OF ARSENICAL BY-PRODUCTS FROM CHEMICAL WARFARE DESTRUCTION EFFLUENTS

T. GUENEGOU, A. JARDY, M. CAUDE.
Laboratoire Environnement et Chimie Analytique, Ecole Supérieure de *Physique et Chimie Industrielles de la ville de Paris, 10 rue Vauquel*in, *75231 Paris Cedex 05 (France).*

A. TAMBUTE
Direction des Centres d'Expertise et d'Essais, E.T.C.- 4, Centre d'Etudes du Bouchet, BP 3, Le Bouchet, 91710 Vert-Le-Petit (France).

1. Summary

Within the context of the chemical warfare destruction (April, 29th 1997 convention), the removal of arsenic, which is highly toxic, from the lewisite hydrolysis effluents has been studied. A process in two steps is presented, which is compatible with industrial and economical constraints.

An efficient way to eliminate the major part of arsenic consists in precipitating it into an insoluble form (calcium arsenate, ferric arsenate or arsenite, etc...). Indeed, the elimination of arsenite and arsenate ions at optimal pH (pH 9 for As^{III} and 4.5 for As^{V}) during the precipitation of ferric hydroxide yields very good elimination rates (> 99.9 %). A molar ratio Fe(III)/As(III or V) as low as 1 is sufficient to get an elimination rate higher than 99%. In addition, this method allows to obtain a good confining of arsenic in the precipitate; for an addition of iron(III) in a molar ratio Fe(III)/As(III or V) equal to 1, the molar ratio As/Fe into the precipitate is also equal to 1. In this way the solid waste volume is minimized.

However, the residual concentration of arsenic is unfortunately greater than the arsenic's maximum contaminant level (M.C.L.). For instance, this one reaches (10.8 ± 0.2) ppm for As^{III} at pH 9 and (1.4 ± 0.1) ppm for As^{V} at pH 4.5. In both cases, iron(III) is added so that the molar ratio Fe(III)/As(III or V) is equal to 1.5. A second step is therefore necessary.

After the batch treatment, the arsenite or arsenate traces can be eliminated by selective adsorption on iron(III) hydroxide precipitated within the pores of a macroporous cation-exchange resin; with a 13 cm x 1 cm i.d column (bed volume equal to 10 ml), an effluent volume equal to 2.8 liters, where the arsenic content is lower than the maximal permissible concentration, is obtained. An extrapolation from these results shows a resin bed volume of 3.6 liters (for instance, a 72 cm x 8 cm i.d. column) would be required to decrease the residual arsenic content of 1 m^3 of effluent from 10 to 0.1 ppm.

F.W. Holm (ed.), Effluents from Alternative Demilitarization Technologies, 85–101.

2. Introduction

The entry into force of the convention (29th April 1997) entails to dispose of efficient and safe methods for the destruction of chemical warfare as simple as possible.

For a moment our laboratory was engaged in studies for the destruction of chemical agents containing arsenic such: Lewisite I and II ($ClCH=CH-As-Cl_2$ (LI) and $(ClCH=CH)_2-As-Cl$ (LII)). The destruction of these compounds don't exhibit peculiar difficulties, indeed these ones are easily and quickly destroyed by reaction in an alkaline medium (aqueous sodium hydroxide) with production of acetylene, sodium chloride and sodium arsenite. However at this stage two difficulties remained:

- The confining of arsenic remains very low and should entail concerns of stockpiling.
- Arsenite ions are highly toxic and safety regulations require that the arsenic concentration in the liquid effluents never exceed legal permissible levels.

If the arsenic's M.C.L. in the liquid wastes has been fixed to 100 $\mu g.l^{-1}$ by the French Legislation [1], this one is reduced to 50 $\mu g.l^{-1}$ for the United States [2,3] and Japan [3]. Consequently, the arsenic content in the post treatment effluents must be in agreement with these standards.

In view of the important content in arsenic (0.1 $mol.l^{-1}$) of lewisite hydrolysis solutions, the detoxication must be carried out in two steps in order to get effluents where the arsenic content is in agreement with the M.C.L. and a solid wastes volume minimized (economic constraint):

- The first one will consist in eliminating in batch the major part of arsenic by precipitation of an insoluble compound. This process should be cheap and allow to decrease significantly the arsenic content in the solution, and to confine arsenic in a small solid volume in order to reduce the cost of its further storage.
- The second step will be a dynamic treatment of effluents resulting from the first step with a selective stationary phase. The arsenic content in column effluent should be in agreement with the M.C.L. (< 100 ppb).

Many investigations deal with the arsenic precipitation under the form of calcium arsenite [4], calcium arsenate [5-7] and ferric arsenate (the scorodite $FeAsO_4$) [6-10], or the trapping of arsenite and arsenate ions during the hydroxides precipitation (calcium [2], aluminium [2] and more frequently iron(III) [2, 3, 9-17]). The solid phase resulting from the trapping of arsenate ions during the ferric hydroxide precipitation is often called "basic ferric arsenate" (sometimes "ferrihydrite bearing arsenate").

Investigations report too the arsenic elimination by anion exchange [2, 18], or by static (in batch) or dynamic (on column) adsorption of the arsenite and/or arsenate ions on hydrous metal oxides (titanium [19], aluminium [2, 18] and more particularly iron(III) [20-27]). Ferric hydroxide is mixed or precipitated into a porous support material, such that titanium oxide (TiO_2) [23], aluminium oxide (Al_2O_3) [23], silica

(SiO_2) [25] or a polymer [20-22, 24]. Authors [27] showed even that arsenic is adsorbed on diagenetic ferric oxyhydroxide in the sediments contained in some Canadian lakes. Finally, authors [18] have studied activated carbon and ferrous sulfide as adsorbent to remove arsenic in groundwater.

This paper deals with the static elimination of arsenite and arsenate ions with iron(III) and the dynamic elimination (on column) of residual arsenite ions by sorption chromatography with a selective stationary phase.

3. Static elimination of arsenite and arsenate ions with iron(III).

3.1. THEORETICAL PART.

We have chosen to operate the static treatment of lewisite hydrolysis solutions with iron(III) because many investigations about the ferric arsenate precipitation report a very good insolubility for basic ferric arsenate. For instance, for a molar ratio Fe^{III}/As^{V} equal to 2, the arsenic concentration is < 0.5 mg.l^{-1} over the pH range of 3-5 at 25°C [9, 11, 13]. The arsenic solubility was shown to decrease with increasing molar ratio Fe^{III}/As^{V}.

Using the values of equilibrium constants reported in the literature (table I), the variations of the conditional solubility of ferric arsenate in presence of Fe(III) at different concentrations versus pH (figure 1) have been studied (program TOT) in order to explain the mechanism of the arsenate ions elimination with iron(III).

TABLE I. Acidity constants of arsenic acid, stability constants of hydroxide complexes of iron(III) and solubility products of ferric hydroxide and ferric arsenate reported from the literature.

Acidity constants of the arsenic acid [6]:	
$H_3AsO_4 \rightleftharpoons H_2AsO_4^- + H^+$	$pKa_3 = 2,1$
$H_2AsO_4^- \rightleftharpoons HAsO_4^{2-} + H^+$	$pKa_2 = 6,7$
$HAsO_4^{2-} \rightleftharpoons AsO_4^{3-} + H^+$	$pKa_1 = 11,2$
Stability constants of hydroxide complexes of iron(III) [6]:	
$Fe^{3+} + OH^- \rightleftharpoons FeOH^{2+}$	$\log K_f = 11,0$
$Fe^{3+} + 2\ OH^- \rightleftharpoons Fe(OH)_2^+$	$\log K_f = 21,7$
$2\ Fe^{3+} + 2\ OH^- \rightleftharpoons Fe_2(OH)_2^{4+}$	$\log K_f = 25,1$
Solubility products of $FeAsO_4$ and $Fe(OH)_3$:	
$FeAsO_4(s) \rightleftharpoons Fe^{3+} + AsO_4^{3-}$	$pK_s = 20,0$ [6]
$Fe(OH)_3(s) \rightleftharpoons Fe^{3+} + 3\ OH^-$	$pK_s = 37,0$ [7]

We can observe (figure 1) an increase of the ferric arsenate solubility on both sides of minimum resulting, towards acid media, from the formation of hydrogenated species of arsenate ($H_2AsO_4^-$ and H_3AsO_4), and towards basic media, from the preferential precipitation of ferric hydroxide, $Fe(OH)_3(s)$ becoming more insoluble than $FeAsO_4(s)$.

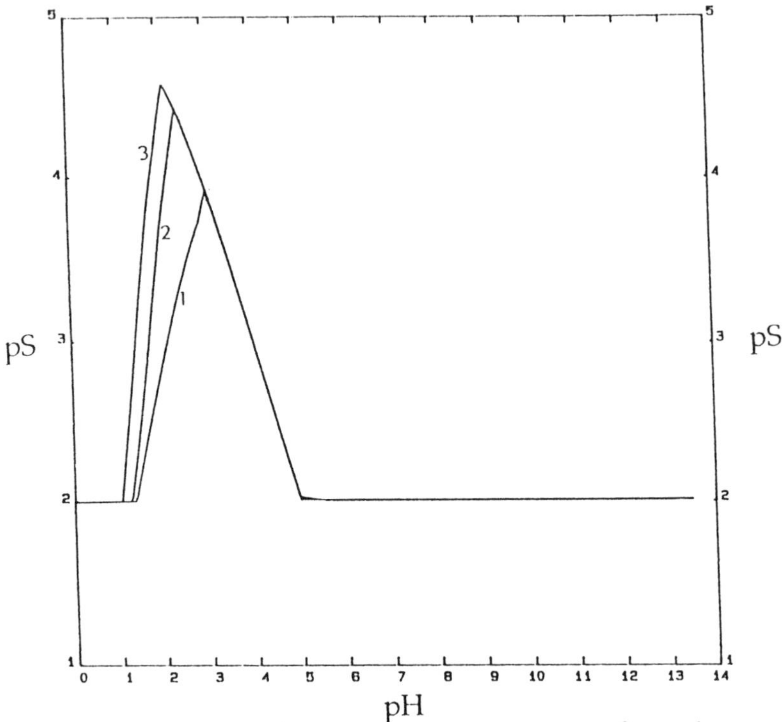

Figure 1: Variation of the ferric arsenate solubility (arsenite concentration: 10^{-2} mol.l^{-1}) in presence of iron(III) at different concentrations (plot 1: 10^{-2} mol.l^{-1}; plot 2: 2.10^{-2} mol.l^{-1}; plot 3: 9.10^{-2} mol.l^{-1}) versus pH.

It appears too (figure 1) that the optimal pH value for the ferric arsenate precipitation depends on the initial molar ratio Fe^{III}/As^{V}. The greater quantity iron(III) the lower the optimal pH yielding a minimum solubility for ferric arsenate.

This last observation is in disagreement with the literature results about ferric arsenate precipitation carried out with initial molar ratios $Fe^{III}/As^{V} > 1$ [2, 3, 9-14]. In this experimental condition, the optimal pH for arsenate ion elimination is less acid than that predicted by theory (figure 1). This discrepancy between the experiment and the theory is due to the co-precipitation of ferric hydroxide, which generates on its surface a complex between ferric ions and arsenate ions (ignored in the theory) resulting from either an ion exchange between the arsenate oxo-anions ($H_iAsO_4^{(3-i)-}$) and the hydroxide ions (OH^-) [7] (unlikely with chlorides 10 times more concentrated than arsenic), or a selective adsorption of arsenate on ferric hydroxide [7, 15-17].

This other mechanism results in a solid phase ("basic ferric arsenate" or "ferrihydrite bearing arsenate") more insoluble than $FeAsO_4(s)$ (the scorodite) and entails operating to a pH less acid that the pH value required for the scorodite precipitation [2, 3, 9-14].

No literature data can be found for the ferric arsenite precipitation; Only an adsorption mechanism of arsenite ions on ferric hydroxide is reported to explain the arsenite retention observed with iron(III) [2, 3].

3.2. EXPERIMENTAL RESULTS

3.2.1 *Ferric arsenate and ferric hydroxide co-precipitation*

pH influence: The variations of the residual arsenate concentration versus the pH for three initial molar ratios Fe^{III}/As^{V} (9, 2 and 1) have been studied (figure 2).

We can observe that the optimal pH value varies with the initial molar ratio. This result is linked to the existence of two mechanisms at the origin of the arsenate ions elimination with iron(III) (cf theoretical part).

For a molar ratio Fe^{III}/As^{V} equal to 9, the arsenate ions elimination can be explained by a mechanism of arsenate selective adsorption on the ferric hydroxide, which leads to the basic ferric arsenate precipitation. The optimal pH value (pH = 5) found for a minimum solubility of the precipitate agrees with that reported in the literature [9, 11-14].

For a molar ratio Fe^{III}/As^{V} equal to 1, the arsenate ions elimination is governed by the ferric arsenate (the amorphous scorodite $FeAsO_4$) precipitation. The optimal pH value (pH = 3) for a minimum solubility of ferric arsenate is in agreement with that predicted by the theory (figure 1).

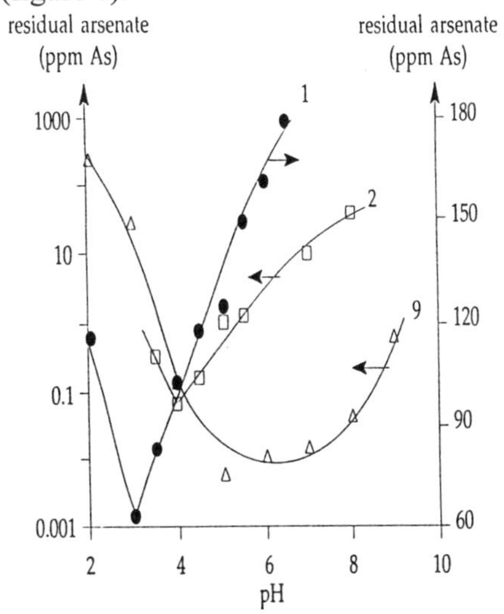

Figure 2: Variation of the residual arsenate concentration versus pH for three initial molar ratio Fe^{III}/As^{V} (9, 2 et 1).

Finally, for a molar ratio Fe^{III}/As^{V} equal to 2, both two mechanisms play a part in the arsenate elimination; the optimal pH value (pH = 4) observed is a compromise between the optimal pH values entailing a minimum solubility of ferric arsenate in case the initial molar ratio Fe^{III}/As^{V} is equal to 1 and 9.

Molar ratio Fe^{III}/As^{V} influence for a pH 4.5. In order to minimize the solid wastes volume (economic constraint), the influence of the initial molar ratio Fe^{III}/As^{V} at pH 4.5 on both the elimination rate of arsenic and the molar ratio As/Fe in the precipitate has been studied (figure 3).

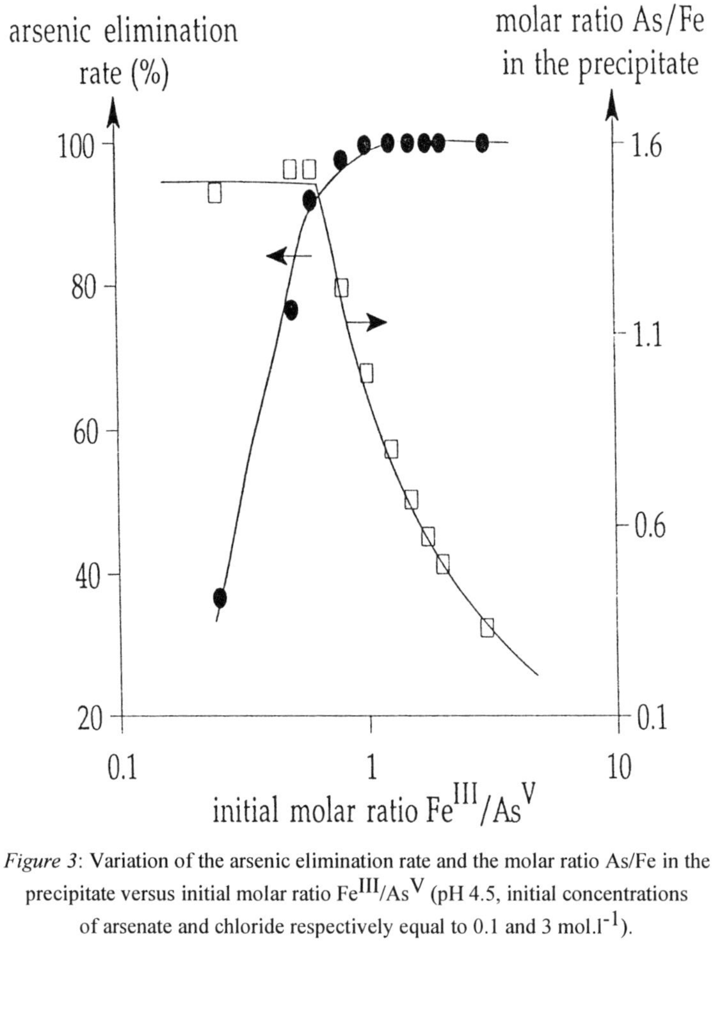

Figure 3: Variation of the arsenic elimination rate and the molar ratio As/Fe in the precipitate versus initial molar ratio Fe^{III}/As^{V} (pH 4.5, initial concentrations of arsenate and chloride respectively equal to 0.1 and 3 $mol.l^{-1}$).

At pH 4.5 and with chloride ions 30 times more concentrated than arsenic, we must operate with an initial molar ratio Fe^{III}/As^{V} slightly greater than 1 in order to obtain both a good confining of arsenic in the precipitate (molar ratio As/Fe   1) and its quantitative elimination of the solution (elimination rate > 99%).

In our experimental conditions, when the initial molar ratio Fe^{III}/As^{V} is   2/3, a define compound with a molar ratio As/Fe equal to 1.5 (for instance $Fe_2(HAsO_4)_3$) is precipitated. This compound has been identified from 1950 [28].

3.2.2. *Arsenite ions trapping during the ferric hydroxide precipitation*
pH influence: The variations of the residual arsenite concentration versus the pH for three initial molar ratios Fe^{III}/As^{III} (10, 2 and 1) have been investigated (figure 4).

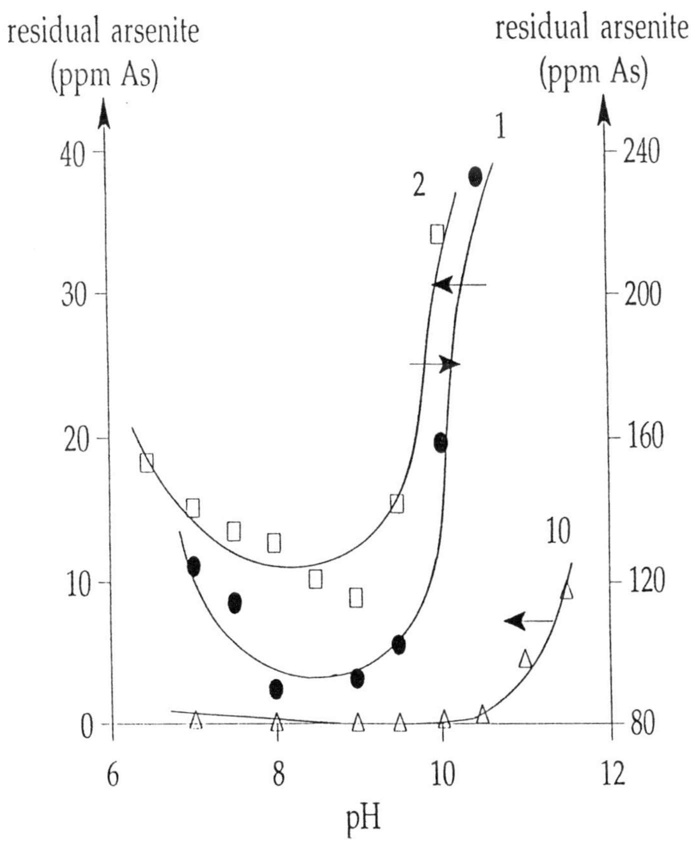

Figure 4: Variation of the residual arsenite concentration versus pH for three initial molar ratio Fe^{III}/As^{III} (10, 2 et 1)

It appears no significant dependence of the optimal pH for arsenite ions elimination with the initial molar ratio Fe^{III}/As^{III}. The optimal pH value (8.5) is in agreement with that reported in the literature [3, 14].

We can notice that the pH required to reach the minimum solubility of basic ferric arsenate is much more acid that the optimal pH for the As^{III} elimination; this latter is much more compatible with the pH of lewisite hydrolysis solutions (the acidity of iron(III) solutions allows to reduce easily the pH from 14 to 9).

Consequently, it is easier to carry out the static treatment of lewisite hydrolysis solutions from As^{III} rather As^{V}.

Molar ratio Fe^{III}/As^{III} *influence at the optimal pH (9)*. The influence of the initial molar ratio Fe^{III}/As^{III} at pH 9 on both the elimination rate of arsenic and the molar ratio As/Fe in the precipitate has been studied (figure 5).

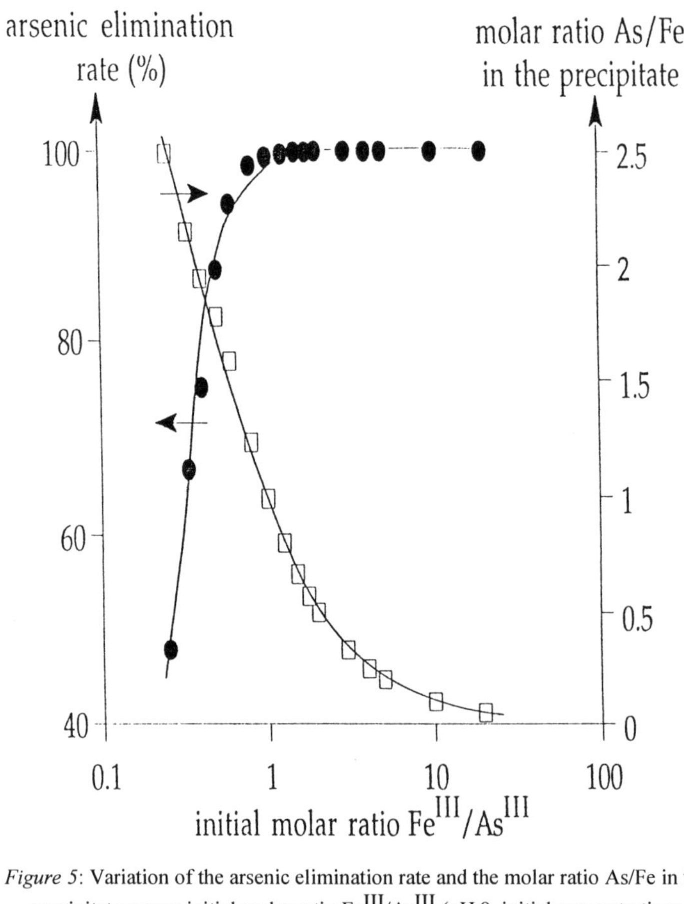

Figure 5: Variation of the arsenic elimination rate and the molar ratio As/Fe in the precipitate versus initial molar ratio Fe^{III}/As^{III} (pH 9, initial concentrations of arsenite and chloride respectively equal to 0.1 and 3 $mol.l^{-1}$).

As in the case of arsenate, we can notice a high arsenic confining in the precipitate (molar ratio As/Fe 1) associated with a quantitative elimination of arsenite (elimination rate > 99%) requires a initial molar ratio Fe^{III}/As^{III} slightly greater than 1.

The decrease of the arsenite elimination rate obtained for the molar ratios lower than 1 is related to a high ferric content in the solution as shown on the figure 6.

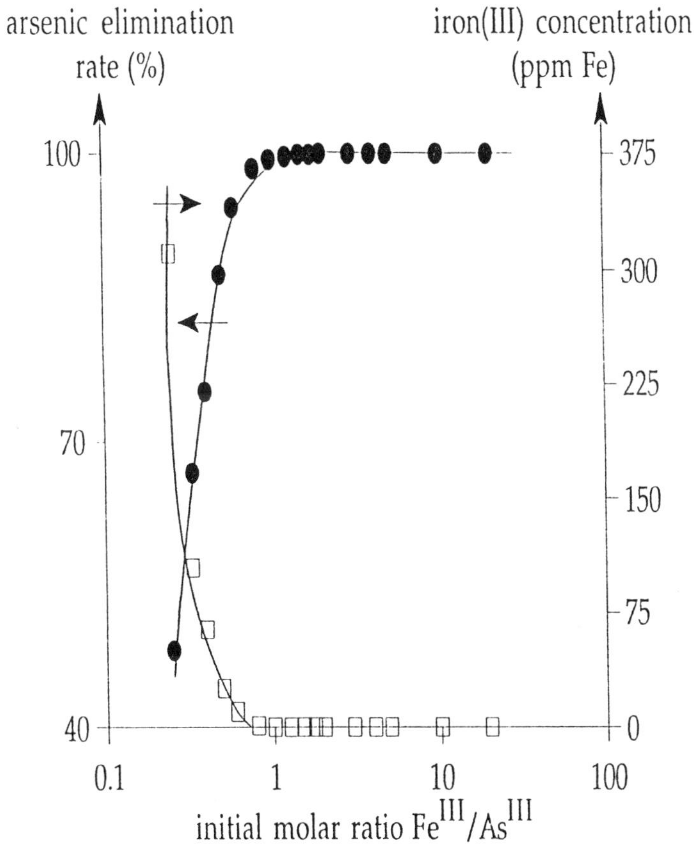

Figure 6: Variation of the arsenic elimination rate and the iron(III) concentration versus initial molar ratio Fe^{III}/As^{III} (experimental conditions: cf figure 5).

4. Dynamic (on-column) elimination of arsenite ions by adsorption chromatography with a selective stationary phase.

4.1 THEORETICAL PART

In view of the high content in chloride ions in the lewisite hydrolysis solutions (10 times more concentrated than arsenic), the final detoxication must be realized with a selective stationary phase; consequently, the use of conventional anion exchangers has not be considered due to their lack of selectivity.

In the table II different adsorbents investigated in the literature are reported. Much of these adsorbents are hydrous metal oxides, and particularly, the ferric hydroxide precipitated on the surface of a porous material used only as support.

Both resins Chelex 100 [21] (a cross-linked polystyrene matrix with iminodiacetic functionality) and DOW XFS 4195 [22] (a cross-linked polystyrene matrix with bis(2-picolylamine) functionality) exhibit the greater saturation As^{III} and As^{V} adsorption

TABLE II. Selective adsorbents for arsenic retention

Adsorbent	capacity (mg As/g)	As (ppm) Influent	As (ppm) effluent	Reference
Resin Dow XFS-4195 used in its hydrolysed Fe(III) form	50	150	<1	22
Resin Chelex 100 used in its hydrolysed Fe(III) form	45* 70**	150	<1	21
Hydrous titanium oxide	1.24	5	0.03	19
Hydrous ferric oxide bonded with a polyacrylamide	23	10^4	500	20
active carbon	1.8	144	135	18
active alumina	12.4	144	82	18
ferrous sulfide	12.6	144	18.4	18
silica gel impregnated with ferric hydroxide	5.6	100	-	25
titanium oxide impregnated with ferric hydroxide	8.7	2.9	-	23
titanium oxide mixed with ferric hydroxide	3.2	2.5	-	23
alumina impregnated with ferric hydroxide	3.5	3.5	-	23
mixture 1:1 Al_2O_3 and TiO_2 each other impregnated with ferric hydroxide	4.2	4.0	,-	23

*: 45 mg As^{V} /g of adsorbent

**: 70 mg As^{III} /g of adsorbent

capacities (about 50 mg As/g resin DOW XFS 4195 (not commercially available) both As^{III} and As^{V}, and respectively 70 and 45 mg As/g resin Chelex 100). These weak base chelating resins have been used in the hydrolysed Fe(III) form.

From [21], the Fe(III) hydrolysis would lead to the substitution of hydration H_2O molecules of the ferric cation by ions OH^-. The study about arsenite and arsenate ions fixation versus the pH (figure 7 [21]) shows that the As^{III} and As^{V} retention is maximal in the pH range where the $H_2AsO_3^-$ and $H_2AsO_4^-$ species predominate (cf figures 8A and 8B).

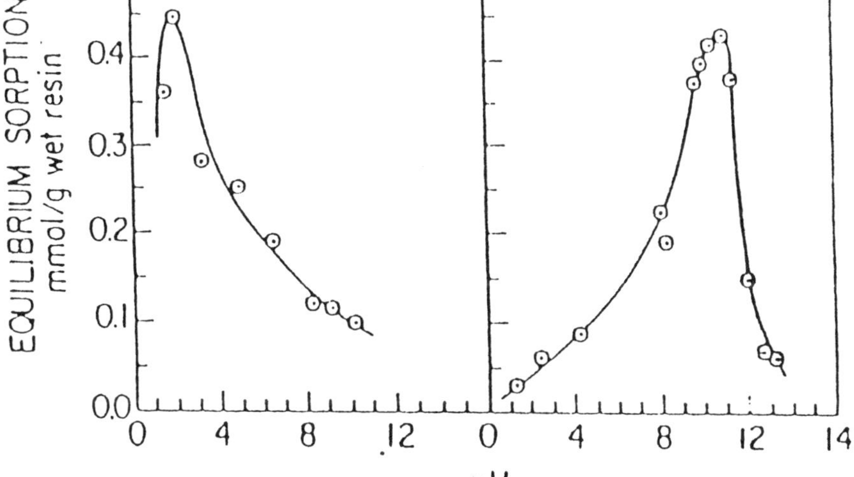

Figure 7: Effect of pH on equilibrium adsorption of arsenate and arsenite by Chelex(Fe^{III}) resin (initial concentration of As^{III} and As^{V} 4 mmol.l^{-1}; resin loading 4 g (wet)/l; temperature 24°C).

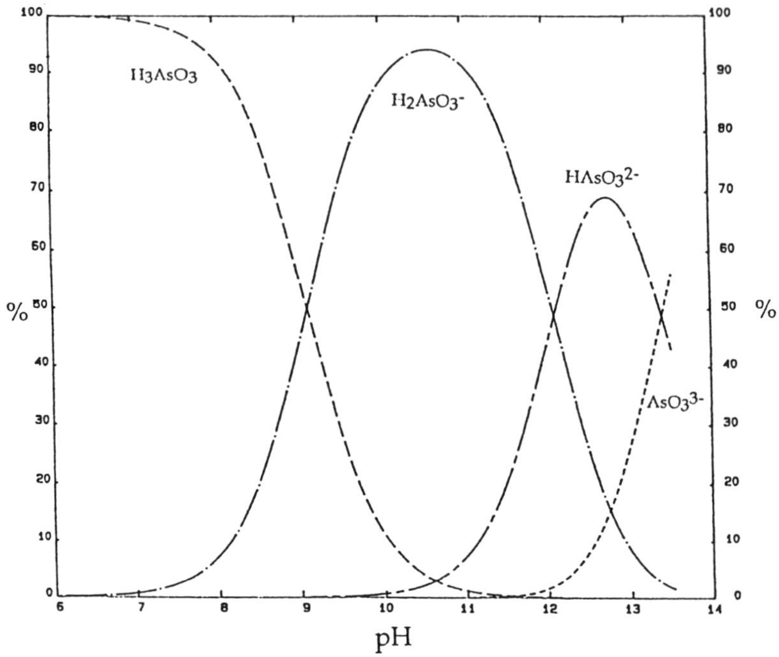

Figure 8a: Distribution diagram of different forms of arsenite versus pH (case of an initial concentration of arsenite lower than the H_3AsO_3(s) solubility).

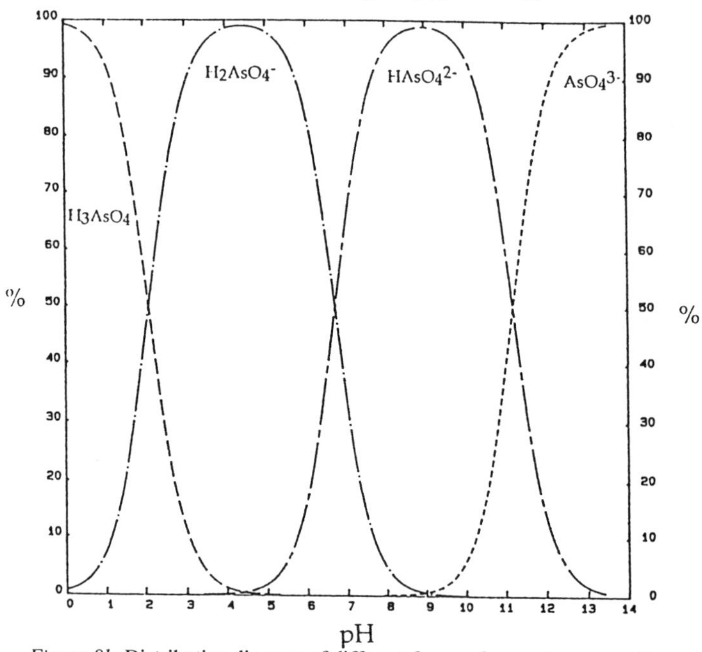

Figure 8b: Distribution diagram of different forms of arsenate versus pH.

Consequently, these authors suggest an exchange of a hydroxide ion (OH^-) with a monovalent anionic specie $H_2AsO_3^-$ or $H_2AsO_4^-$ according to the reaction:

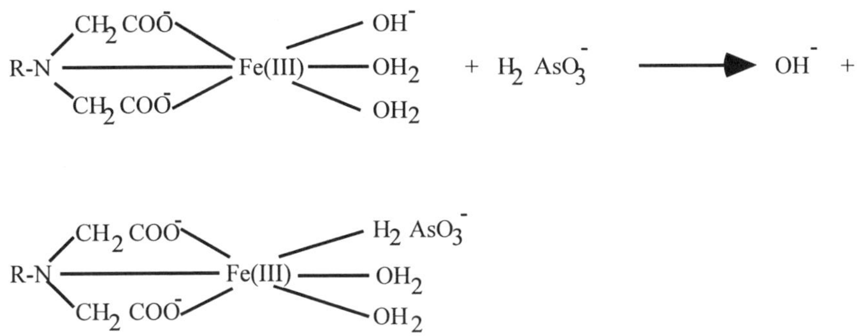

This interpretation can be discussed as for as the fixed arsenic quantity is not modified in the presence of chloride ions of increasing concentration; this later is even increased in the case of As^V [21].

Taking into account the fact another macroporous materials treated in surface with ferric hydroxide (table II) exhibit a retention for arsenite and arsenate ions, an another explanation can be found in the adsorption of these latters on the colloidal ferric hydroxide precipitated into the pores of adsorbents.

4.2. EXPERIMENTAL RESULTS.

The ferric hydroxide precipitation into the chelating resin Chelex 100 entails very quickly a large decrease of the flow rate (for a constant pressure drop) due to a decrease of the column permeability. Consequently a macroporous resin have been chosen (BioRad AGMP-50) (the presence of macropores with a diameter of several 10 nm makes easier the access of mobile phase even after the ferric hydroxide precipitation).

The counter-ions of sulphonate groups of the macroporous cation resin are exchanged with ferric ions, then iron(III) is hydrolysed with a sodium hydroxide solution. Thus, ferric hydroxide is scattered into the macropores of the cation exchanger used therefore under its hydrolysed Fe(III) form.

4.2.1 *Influence of chloride ions on the arsenite breakthrough curve.* We have plotted the arsenite breakthrough curves at pH 10.65 with or without chloride ions for a 50 x 1 i.d cm column packed with the macroporous cation exchanger used under its hydrolysed Fe(III) form (figure 9).

The adsorbed arsenic quantity is not significantly affected by the presence of chloride ions one hundred times more concentrated than arsenite ions (370 versus 412 mg As without chloride). Consequently, this result confirm that the mechanism involved in the As^{III} retention is much more complicated than a simple ion exchange mechanism.

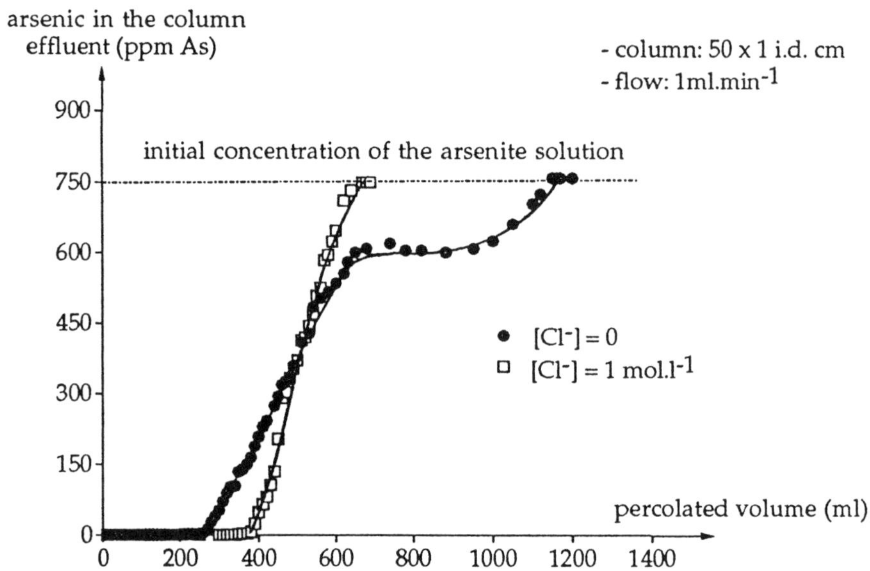

Figure 9: Arsenite breakthrough curves on a macroporous cation exchanger used under its hydrolyzed Fe(III) form of arsenite solutions 10^{-2} mol.l^{-1} (ph10.5) with or without chloride ions.

4.2.2 *Application to supernatants from static treatment with iron(III) of arsenite solutions.* Figure 10 shows the breakthrough curve of supernatants from a static treatment with iron(III) of arsenite solutions (initial concentrations in Cl^- and As^{III} equal respectively to 3 and 0.1 mol.l^{-1}) on a 13 x 1 i.d. cm column packed with the macroporous cationic exchanger used under its hydrolysed Fe(III) form.

The static treatment was carried out at pH 9 (pH entailing an optimal fixation of As^{III} on ferric hydroxide) and with a molar ratio Fe^{III}/As^{III} equal to 1.5 in order to get an arsenite elimination rate >99% (residual arsenite content equal to (10.8 ± 0.2) ppm As) and a good confining of arsenic in the precipitate.

A bed volume equal to 10 ml (13 x 1 i.d. cm column) allows to obtain an effluent volume equal to 2.8 liters, where the arsenic content is lower than 0.1 ppm (the arsenic's M.C.L.).

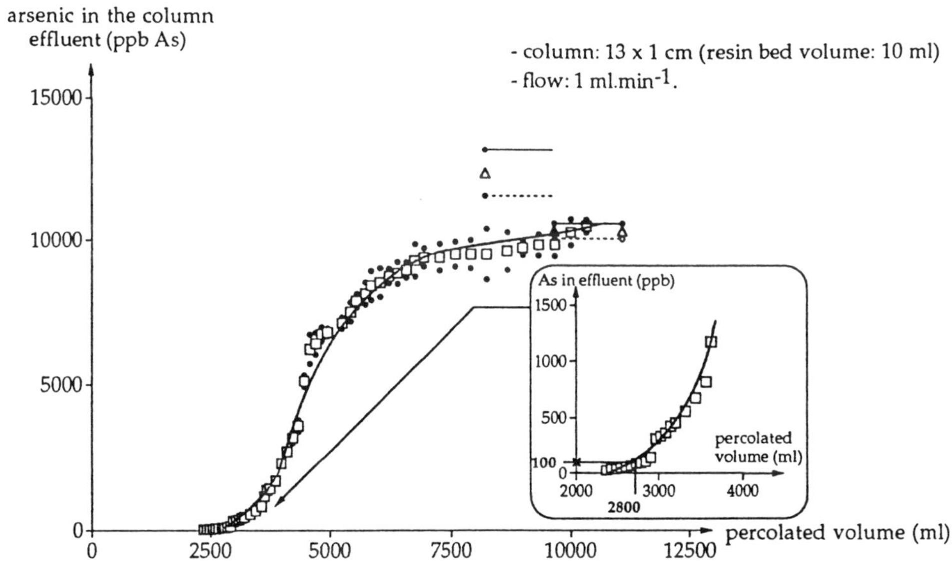

Figure 10: Arsenite breakthrough curve on a macroporous cation exchange used under its hydrolyzed Fe(III) form of arsenite solutions resulting from batch treatment with iron (III) (molar ratio Fe^{III}/As^{III} equal to 1.5, pH 9). The chloride and arsenite initial concentrations are respectively 3 and 0.1 $mol.l^{-1}$

5. Conclusion

The detoxication of lewisite hydrolysis solutions containing arsenite or arsenate ions at concentrations reaching 0.1 $mol.l^{-1}$ should be carried out in two steps for minimizing the solid wastes volume: precipitation of the major part of arsenic (> 99%) under an insoluble form followed by on-column treatment of supernatants with a selective stationary phase.

In the first step a good arsenic confining in the precipitate (As/Fe 1) does not allow to reach a residual arsenic content in the supernatant satisfying the arsenic's

M.C.L. (0.1 ppm).

A residual arsenic concentration lower than 50 ppm is obtained when the arsenite or arsenate ions trapping during the ferric hydroxide precipitation is carried out at pH 9 for As^{III} and at pH 4.5 for As^{V}, with iron(III) added in a molar ratio Fe/As equal to 1 and with chlorides 30 times more concentrated than arsenic.

In the second step, the dynamic treatment of supernatant resulting from the first step is carried out by sorption chromatography with a macroporous cationic exchanger used in its hydrolysed Fe(III) form.

With a resin bed volume equal to 10 ml (13 x 1 i.d. cm column), 2.8 liters effluent volume, in which the arsenic content was in agreement with the M.C.L, is collected.

The extrapolation of these results shows a 3.6 liters resin bed volume (for instance a 72 x 8 i.d. cm column) allows the complete detoxication of 1 m^3 of lewisite hydrolysis solution.

6. References

1. Ministerial order of 18 February 1994, Storage of special industrial wastes, J. O., 1994, 6117-6124.
2. Kartinen E.O. Jr, Martin C.J. (1995) An overview of arsenic removal processes, *Desalination*, 103 (1-2), 79-88.
3. Hering J.G., Chen P.Y., Wilkie J.A., Elimelech M., Liang S. (1996) Arsenic removal by ferric chloride, *J Am Water Works Assoc.*, 88 (4), 155-67.
4. Story L.G., Anderson E. (1924) The action of arsenic trioxide in water solution on certain metallic hydroxides, *J Am Chem Soc*, , 46 (1), 533-538.
5. Robins R.G. (1981) The solubility of metal arsenates, *Metallurgical Transactions B*, 12B, 103.
6. Ringbom A. (1967) *Les complexes en chimie analytique*, Dunod, Paris.
7. Charlot G. (1967, 1983) *Les réactions chimiques en solution aqueuse*, Masson, Paris.
8. Dutrizac J.E., Jambor J.L. (1988) The synthesis of cristalline scorodite $FeAsO_4.2H_2O$, *Hydrometallurgy,* 19, 377-384.
9. Krause E., Ettel V.A. (1989) Solubilities and stabilities of ferric arsenate compounds, *Hydrometallurgy*, 22, 311-337.
10. Demopoulos G.P., Droppert D.J., Van Weert G. (1995) Precipitation of crystalline scorodite ($FeAsO_4.2H_2O$) from chloride solutions, *Hydrometallurgy*, 38, 245-261.
11. Papassiopi N., Vircikova E., Nenov V., Kontopoulos A., Molnar L. (1996) Removal and fixation of arsenic in the form of ferric arsenates. Three parallel experimental studies, *Hydrometallurgy*, 41, 243-253.
12. Nenov V.V., Dimitrova N., Dobrevski I., Rands D.G. (1992) Effective precipitation of arsenic from aqueous solution by iron(III) sulfate, *Acta Hydrochim Hydrobiol.*, 20 (1), 14-17.
13. Vircikova. E., Molnar L., Lech P., Reitznerova E. (1995) Solubilities of amorphous Fe-As precipitates, *Hydrometallurgy*, 38, 111-123.
14. Nishimura T., Wang Q., Umetsu Y. Removal of arsenic from process liquors by oxidation of iron(II), arsenic(III) and sulfur(IV) with oxygen, 535-548.
15. Dutrizac J.E., Harris G.B. (1996) *Iron control and disposal*, Proceedings Of The Second International Symposium On Iron Control In Hydrometallurgy, Ottawa, Canada, October 20-23.
16. Waychunas G.A., Rea B.A., Fuller C.C., Davis J.A. (1993) Surface chemistry of ferrihydrite: Part 1. EXAFS studies of the geometry of coprecipitated and adsorbed arsenate, *Geochim Cosmochim Acta.*, 57 (10), 2251-69.
17. Robins R.G., Wong P.L.M., Nishimura T., Khoe G.H., Huang J.C.Y. (1992) Basic Ferric Arsenates - Non Existent, 31-9.
18. Hager J.P., *EPD Congr. 1992,* Proc. Symp. TMS Annu. Meet., Miner. Met. Mater. Soc. (publisher), Warrendale.
19. Hsia T.H., Lo S.L., Lin C.F. (1992) Arsenic(V) Adsorption On Amorphous Iron Oxide: Triple Layer Modelling, *Chemosphere*, 25 (12), 1825-37.

20. Kuhlmeier P.D., Sherwood S.P. (1996) Treatability of inorganic arsenic and organoarsenicals in groundwater, *Water Environ. Res.*, 68 (5), 946-951.
21. Sato H., Shiegeta S., Uchida H. (1980) Process of treatment of waters containing arsenic and silica, japanese patent n° 55.8843.
22. Shiegetomi Y., Hori Y., Kojima T. (1980) The removal of arsenate in waste water with an adsorbent prepared by binding hydrous iron(III) oxide with polyacrylamide, *Bull. Chem. Soc. Jpn..*, 53, 1475-1476.
23. Chanda M., O'driscoll K.F., Rempel G.L. (1988) Ligand exchange sorption of arsenate and arsenite anions by chelating resins in ferric ion form: iminodiacetic chelating resin Chelex 100, *Reactive Polymers*, 8, 85-95.
24. Chanda M., O'driscoll K.F., Rempel G.L. (1988) Ligand exchange sorption of arsenate and arsenite anions by chelating resins in ferric ion form: weak-base chelating resin DOW XFS-4195, *Reactive Polymers*, 7, 251-261.
25. Hlavay J., Foldi-Polyak K., Inczedy J. (1984) Removal of arsenic from natural waters, *Stud. Environ. Sci.*, 23, 373-380.
26. Isaacson A.E., Corwin R.R., Jeffers T.H. (1994) Arsenic removal using immobilized ferric oxyhydroxides, 47-55.
27. Harris B., Krause E. (1994) *Impurity Control Disposal Hydrometall. Processes*, 24th Annu. Hydrometall. Meet., Can. Inst. Min. Metall. Pet., Montreal, Canada.
28. Yoshida I., Kobayashi H., Ueno K. (1976) Selective adsorption of arsenic ions on silica gel impregnated with ferric hydroxide, *Anal. Lett..*, 9 (12), 1125-1133.
29. Maeda S., Ohki A., Tsurusaki Y., Takeshita T. (1990) Iron(III) hydroxide-loaded coral limestone as an adsorbent fer arsenic(III) and arsenic(V), *Sep. Sci. Technol.*, 25 (5), 547-55.
30. De Vitre R., Belzile N., Tessier A. (1991) Speciation and adsorption of arsenic on diagenetic iron oxyhydroxides, *Limnol. Oceanogr.*, 36 (7), 1480-1485.
31. Nishimura T., Robins R.G. (1996) Crystalline phases in the system Fe(III)-As(V)-H2O at 25°C, 521-534.
32. Dutrizac J.E., Harris G.B. (1996)*Iron control and disposal*, Proceedings Of The Second International Symposium On Iron Control In Hydrometallurgy, Ottawa, Canada, October 20-23.

EFFLUENTS FROM ALTERNATIVE DEMILITARIZATION TECHNOLOGIES

Products of Incomplete Reaction

W. G. MAY
University of Illinois
Chemical Engineering Department
600 S. Mathews Ave.
Urbana, Illinois 61801-3792

1. Products from Alternative Technologies

Chemical weapons are being destroyed by incineration at two storage sites in the USA; the process is referred to as the "base-line technology". Several alternatives are being investigated. Two are at a relatively advanced stage and will be used, one for destroying VX (Nerve Agent) and the other for HD (Mustard). Both are multi-step processes in which the chemical agent is first detoxified (neutralized) and then treated further to produce products suitable for discharge. These two processes will be examined in this paper, and their products compared with those of the base-line technology.

2. Basis for Comparison

Any new destruction technology should be compared with current practice, i.e. base-line incineration. The base-line consists of combustion (with added fuel), followed by caustic scrubbing to remove acid gases; the scrubbing solution is evaporated to dryness. The products from the agent destruction which leave the plant are: flue gas, dried salts, and steam from the evaporator.

Toxic waste incinerators are required to demonstrate performance through initial "trial burns", involving very detailed analysis of the products. This establishes operating parameters for the incinerator for subsequent control. It is assumed that if the operation is held within the established parameters - combustion chamber temperature, percent excess oxygen, residence time (i.e. throughput), carbon monoxide level in the flue gas, - the operation will match the trial burn. In the case of the U.S. Army's incinerators (at Johnston Island, and at Tooele, Utah) an additional operating parameter is required: the flue gas is monitored for unburned agent; the plant is shut down if any is detected.

Analytical data from trial burns at the Johnston Island incinerator are shown in Table 1. Various products have been identified in the stack gas from incineration of

F.W. Holm (ed.), Effluents from Alternative Demilitarization Technologies, 103–120.

three agents: VX and GB nerve agents and mustard. Table 1 is a sampling of them. There are many more in the original data. Additional fuel, diesel oil, was also being burned.

No agent was detected in the flue gas. The ASC (acceptable stack concentration) has been set at 0.0003 mg/m^3, (roughly 300 parts per trillion on a weight basis). To ensure meeting this limit the analytical instrument has a DL (detection limit) of 60 parts per trillion; thus the unburned agent remaining in the flue gas was at a level of 60 ppt or less. The plant owner (U. S. Army) has also placed a requirement on the incinerator of at least 99.9999% (6-nines) destruction of agent. In this case the residual level of less than 60 ppt implies a destruction of 8-nines.

Chlorinated dioxins and furans are of particular concern and are subject to a particularly exacting analysis. The detection limits shown in Table 1 are in the range of 0.1 to l ppt.

All incinerator flue gases contain a large number of PICs (products of incomplete combustion). They are not considered as hazardous as chemical agent or chlorinated dioxins/furans. They are present at very low levels, though much higher than the agent or dioxins/furans; they are present in the range of parts per billion to parts per million (ppb to ppm). (Some that show up in the analysis are believed to be artifacts of the analytical procedure, rather than materials originally in the flue gas.)

In summary, combustion is capable of a very high degree of "mineralization" when properly carried out. A large number of materials have been identified in the stack gas but at very low concentration, many at less than ppb levels. Materials of most concern, unburned agent and chlorinated dioxins/furans, are at parts per trillion levels.

The other two residual streams from the combustion process - the recovered salts and the steam from the evaporator - are free of residual agent. They have not been examined in detail for other organics but should have very low levels, in line with the low levels of PICs in the flue gas.

3. Chemical Neutralization: VX

It has been proposed that ton-containers of VX at one storage site should be chemically neutralized, rather than incinerated. The hydrolyzed product of the reaction, an aqueous mixture of organic materials with excess NaOH (the hydrolysis agent) will then be treated by super critical water oxidation (SCWO). The data for the hydrolysis product are of interest since they are the feed to the SCWO; the ultimate products from the process however come for the most part from the SCWO unit.

The major reactions occurring in the neutralization process are shown in Fig. 1.

- Approximately 90% of the VX reacts directly (by the top reaction) to form ethoxy-methyl phosphonic acid (EMPA) or its sodium salt, and the bis (propyl) amino ethyl thiol ("thiolamine").

- Part of the VX reacts (about 10%) initially to form the intermediate compound shown as "EA2192". This is itself highly toxic, but will react slowly to form methyl phosphonic acid (MPA) and more "Thiol".

The reaction requires vigorous mixing since VX is quite insoluble in water. Most of the products are soluble, but a small residue, arising mainly from a stabilizer

TABLE 1
Johnson Island Incinerators
Trial Burns for VX, GB, HD

Emissions of Substances of Potential Concern (SOPC)

SOPC	Regulated Pollutant	Maximum Stack Gas Line ug/m^3 (approx. ppm weight)	Detection Limit (D_2)
VX*	Yes	6.0×10^{-2}	6.0×10^{-2}
GB*	Yes	6.0×10^{-2}	6.0×10^{-2}
HD*	Yes	8.7	8.7
PCDD/PCDF Congeners			
2, 3, 7, 8 tetrachloro di benzo-p-Dioxin	Yes	8.83×10^{-5}	1.10×10^{-5}
Pentachloro DBD*	Yes	5.66×10^{-5}	5.66×10^{-5}
2378 tetra chloro di benzo furan	Yes	1.16×10^{-5}	7.58×10^{-6}
Pentachloro DBF	Yes	1.34×10^{-4}	4.23×10^{-5}
Volatile PICs			
Benzene	Yes	4.75×10^{1}	3.0×10^{-1}
2 Butanone	No	1.47×10^{2}	1.0×10^{-1}
Carbondisulfide	No	1.0×10^{1}	1.0×10^{-1}
Chloroform	No	3.96×10^{1}	1.0×10^{-1}
Styrene	No	2.41×10^{2}	1.0×10^{-1}
Vinyl chloride	Yes	9.92	1.0×10^{-1}
Semi Volatile PICs			
Benzoic Acid	No	1.23×10^{2}	1.40
2-Methyl phenol	No	1.24×10^{2}	1.40
Naphthalene	No	2.92	1.40
Pollutants			
Particulates	Yes	8.0×10^{3}	
PCB	Yes	8.5×10^{-2}	
Metals			
Antimony*	Yes	1.6 x 10	1.6 x 10
Arsenic	Yes	2.2 x 10	6.78
Chromium	Yes	1.45 x 10	3.75
Copper	Yes	8.7	7.5
Mercury	Yes	1.01 x 10	5.36
Vanadium	Yes	1.07 x 10	2.56
Zinc	Yes	2.31×10^{2}	6.75

* Undetected. The maximum stack gas concentration was assumed equal to the detection limit in cases where the contaminant, if regulated, was not detected.

added to the original VX, remains as an insoluble organic layer on the water. The reaction is carried out in a batch reactor; a reaction time of about 6 hours at a temperature of 90°C is required, primarily to eliminate the intermediate "EA2192".

Product analyses for the hydrolysis part of the process will focus on the products shown in Fig. 1.

The VX has a purity level of 90-95%; a typical analysis is shown in Table 2. GC obtained the agent analysis with a thermal conductivity detector. The other organic components were analyzed by GC and Electron-impact mass selective detector. Metal analysis used inductively coupled plasma/atomic emission spectroscopy (ICP/AES). Mercury was by cold vapor atomic absorption.

The largest "impurity" listed is diisopropylcarbodiimide, added to the VX as a stabilizer (anti oxidant). It can be as much as 5%. The other impurities shown are present at large "ppm" concentrations; many more would probably be found with more sensitive analyses - e.g. in the "ppb" range.

The hydrolysate product has two phases: an aqueous phase containing most of the product and an organic layer with approximately 5 percent of the total organic. The compositions of the two-product phases will of course, be quite different. A typical composition for the mixture (obtained by "homogenizing" the two phases) is shown in Table 3. The analysis was based on NMR of ^{13}C for the carbon species, ^{31}P for the phosphorous species.

The major products are those expected from the reaction sequence of Fig. 1, together with non-reacted impurities expected from the initial composition, Table 2. Materials present at very low concentrations e.g. ppm range or less are not, in general,

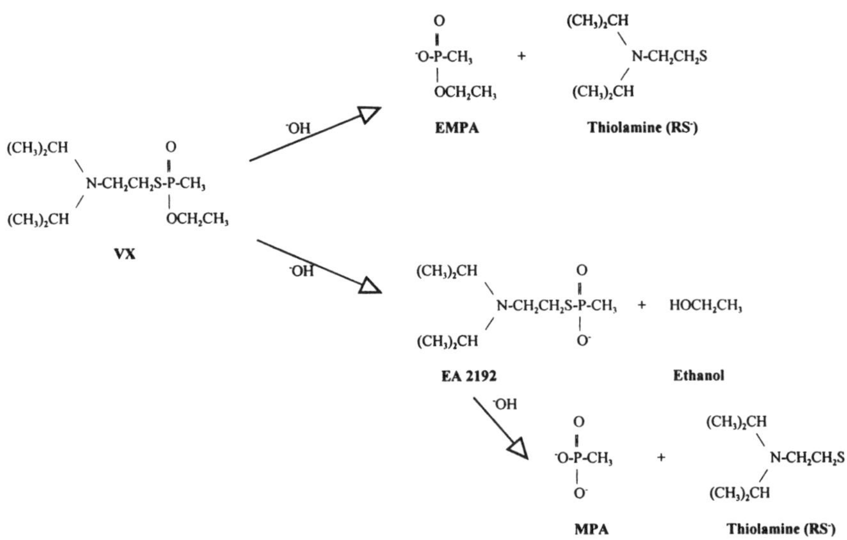

Figure 1. VX Neutralization, Primary Reactions

TABLE 2: VX TON CONTAINER SURVEY

Organic Compounds	Analytical Results (Area %)	Normalized Mole%
VX by GC/TCD	92.7	92.7
Dimethyl Ketone (Acetone)	0.009	0.013
Diisopropylamine	0.15	0.22
N, N-Diisopropylmethylamine		NR
Diisopropylcarbodiimide (Stabilizer)	3.1	4.6
Unknown (-2.3)		NR
N, N-Diisopropylethylamine		NR
Unknown (3.5 to 3.8)		NR
O-Ethyl methylethylphosphinate		NR
1, 3-Diisopropylurea	0.065	0.095
Diethyl methylphosphonate	0.031	0.045
2-(Diisopropylamino)ethane thiol	0.70	1.01
Unknown (6.8 to 7.1)		NR
Unknown (7..3 to 7.7)	0.098	0.143
O, O-Diethyl methylphosphonothioate	0.130	0.19
O, S-Diethyl methylphosphonothiolate	0.030	0.044
2-(Diisopropylamino)ethyl ethyl sulfide	0.055	0.080
Unknown (7.9 to 8.1)	0.049	0.071
Unknown (8.6 to 9.0)	0.050	0.073
Diethyl Dimethylpyrophosphonate (Pyro)	0.085	0.124
O, O-Diethyl dimethylpyrophosphonothioate	0.026	0.038
O-(2-Diisopropylaminoethyl) O-ethylmethylphosphonate (QB)	0.077	0.112
Unknown (-13.5)		NR
1, 2-bis (ethyl methylphosphonothiolo)ethane	0.31	0.45
Unknown (14.2 to 14.4)		NR
Unknown (-14.9)		NR
Unknown (16.1 to 16.5)		NR
Unknown (16.9 to 17.2)		NR
Unknown (-18.6)		NR
Unknown (18.9 to 19.1)		NR
Unknown (- 19.9)		NR
SUMS	4.99	100.00
		Impurity Factor = 1.46
Metals	Analytical Result (ug/g)	
Phosphorus (wt. %)	11.5 %	
Arsenic	ND	
Chromium	ND	
Lead	ND	
Selenium	ND	
Zinc	2.5	
Mercury	0.47	
Copper	ND	
Iron	6.9	
Magnesium	ND	
Calcium	ND	
Barium	ND	
Sulfur (wt. %)	10.9 %	
Silicon	ND	

Wt. % of Phosphorus should be 11% +/-0.25% & wt. % of Sulfur should be 11.4 +/-0.2% based on VX purity only. NR = Not Reported. ND = Not Detected

TABLE 3
Analysis of Homogenized VX Hydrolysate
240 - Minute Reaction Time

Analysis	Results: Compound	Mol %
^{13}C and ^{31}P Combined NMR Results	EMPA	41.5
	MPA	4.6
	EMPSH	0.3
	Other ^{31}P	0.2
	RSH	43.4
	RSSR	0.4
	RSR	0.7
	RSCCSH	1.3
	Ethanol	2.3
	DIPA	0.6
	CDI/products	2.7
	Other RS-compounds	1.9

COMMON ABBREVIATIONS, NAMES AND STRUCTURES USED BY NMR

Abbreviations	Chemical Name/Class	Structure
EMPA	Ethyl methylphosphonic acid	$OP(CH_3)(OCH_2CH_3)(OH)$
MPA	Methylphosphonic acid	$OP(CH_3)(OH)_2$
EMPSH	O-Ethyl methylphosphonothioic acid	$SP(CH_3)(OCH_2CH_3)(OH)$
RSH	Diisopropylaminoethanethiol	R'- SH
RSSR	Diisopropylaminoethane disulfide	R'- SS'R'
RSR	Diisopropylaminoethane sulfide	R' - S - R'
RSCCSH	Diisopropylaminoethylthioethy sulfide	R' - S' CH_2CH_2SH
DIPA	Diisopropylamine	$HN[CH(CH_3)_2]_2$
CDI	Dicyclohexylcarbodiimide	$C6H_5 - N = C = N - 6H_5$

differentiated in the analysis. The exception is VX, shown to be reacted to the part per billion range, Table 4.

TABLE 4
12-Litre Test Reactor
Agent and EA 2192 Analysis

Reaction Time Minutes	Homogenized Samples: VX *	Homogenized Samples: EA 2192 *
30	ND < 20 ppb	65 ppm
120	ND < 20 ppb	ND < 24 ppm

* GC/ITMS analysis
NMR P^{31} analysis

The primary purpose of the VX-hydrolysis is to reduce the material toxicity. The products mixture must be treated further (mineralized) before release. Animal tests have been carried out to confirm the reduced toxicity. The intravenous LD 50 value of VX for mice was reported to be 0.0141 mg/kg. The hydrolysis product was 4 to 6 orders of magnitude less toxic. Typical results are shown in Table 5.

TABLE 5
Mouse Intravenous LD50/Toxcity Data VX Reacted with Caustic (NaOH)

Sample	LD50 mg/kg 24Hr	Toxic Signs
Homogenized Material	349 (299 - 420)	Ataxia, Convulsive jerks, exophthalmus, collapse, prostration death
Top Layer	71.5 (65.1 - 78.6)	Ataxia, Convulsive jerks, exophtalmus, collapse, prostration death
Mid Layer	2024 (1840 - 2198)	Ataxia, Convulsive jerks, exophtalmus, collapse, prostration death
"Pure" VX	0.0141	

The characteristic signs of nerve agent attack were not observed with the hydrolyzed material. The toxicity is due to the salt and organic content of the samples.

In summary, the neutralization of VX with NaOH reduces the nerve agent and its related toxicity to very low levels. The technology chosen for treatment to release is super critical water oxidation.

4. Supercritical Water Oxidation of VX Hydrolysate

There are several streams fed to the SCWO reactor.

- The main organic stream is the hydrolysate from the neutralization unit. It has a composition of about 34% organic material from the original VX, the balance water and caustic;
- Two other aqueous streams contain a small amount of hydrolyzed VX: A dilute hot caustic stream arising from the clean-out of the ton-containers after the agent VX was drained; and any decontamination fluid (solution of NaOH) used for clean-up, e.g. of "proximity suits" worn for direct access to agent. These two additional aqueous streams will dilute the organics to about 20% of the liquid.
- Other streams consist of an oxidant and water. The oxidant used in pilot testing has been air, generally at 100% excess over stoichiometric. The oxidant planned for the final plant is liquid oxygen at 50% excess. The hydrolysate has sufficient organic to raise the reactor temperature above the desired level (about 650°C); water will be added for reactor temperature control.

Products from the SCWO unit are: A gas stream consisting of excess oxygen plus some products of the oxidation (e.g. CO, CO_2 N-oxide,); and an aqueous solution of salts formed by the oxidation process, e.g. Na_3PO_4, Na_2SO_4.

An analysis of the gas stream is shown in Table 6. The oxidizer used in this experimental test was air so the exit gas had a large nitrogen content. The plant operation based on liquid oxygen would have very little nitrogen. SCWO experimenters have claimed that the nitrogen in VX will appear in the product gas as N_2O. This is not shown in Table 6; no test was done for it.

The proposed process has some similarities to the base-line process. In both cases, the organic material (VX) is mineralized by oxidation, and the acidic components are removed from the gas by contact with caustic-solution. The gas compositions from the two processes appear very different however. The SCWO off-gas contains little of the PICs observed with normal combustion. Only 5 volatile PICs were observed, at concentrations of 9 to 41 ppb, - in contrast to the 50 or more identified in incinerator trial burns. The number observed from SCWO conditions may increase with more intensive analytical observation.

TABLE 6
Gaseous Effluent Analysis Results for 8-Hour VX/NaOH Hydrolysate Test

Sample No.	O_2 (%)	N_2 (%)	CO_2 (%)	CO (ppm)	NO_x (ppm)	SO_2 (ppm)	VOC (ppb)
Baseline(2)	10.8	76.9	7.9	<3	0.35	<8.6	9-16(1)
1(3)	8.8	78.0	8.5	<3	6.9	<9.2	7-28(1)
2(4)	12.4	77.2	6.0	<3	4.0	<9.9	14-41(1)

(1) Analyses were conducted for approximately 50 different volatile organic compounds. Of these, only five were detected, all at very low concentrations. Acetonitrile, acetone, 2-butanone, carbon disulfide, and vinyl acetate were detected in concentrations of 9 to 41 ppb. The most prominent volatile organic was acetone, which was present at 41 ppb in Sample No. 2.

(2) Baseline gas sample collected between 0831 and 0858 hours.

(3) Gas sample No. 1 collected between 1343 and 1405 hours (193 to 215 minutes after the start of steady state).

(4) Gas sample No. 3 collected between 1625 and 1643 (355 to 373 minutes after the start of steady state).

It has been shown in other work that the hydrolytic oxidation of SCWO goes through a series of steps, with some intermediate species being particularly refractory; they limit the overall oxidation rate. The conditions chosen for this work (temperature of 650°C, residence time of ½ minute) are chosen to break the very refractory C-P bond in MPA; they are apparently sufficient to oxidize most other organic species to final products.

Compositions and rates for a large SCWO test are shown in Table 7. The major inlet species are oxidized to four-nines or more, with the exception of the phosphoric acids (EMPA and MPA). The incomplete conversion of these two species is an example of the slow breakdown of the C-P bond. Conversion of MPA (methyl phosphonic acid) appears to be the rate-limiting process. Ultimately the heteroatoms in the original VX (i.e. P and S) are mineralized to their stable salts; the 'N' presumably goes to N_2O.

It is not clear why a material such as acetone will show up in the gas phase analysis (albeit at a low level), but not in the liquid phase. [Gas and Liquid are separated

TABLE 7
Destruction Efficiencies for August 13, 1996 Hydrolysate Test

Comp.	Feed Conc. (ppm)	Inlet rate (kg/hr)	Liquid Eff. Conc. (ppm)	Liquid Eff. Rate (kg/hr)	Gas Eff. Conc. (ppm)	Gas Eff. Rate	Total Eff. Rate (kg/hr)	DRE
RSH	78,300	1.588	<0.008	<1.25E-06	<0.001[1]	<1.56E-07	<1.40E-6	>99.9999%
RSSR	13,300	0.270	<0.008	<1.25E-06	<0.001[1]	<1.56E-07	<1.40E-6	>99.9995%
MHA	7,000	0.142	<0.008	<1.25E-06	<0.001[1]	<1.56E-07	<1.40E-6	>99.9990%
DIAE	5,600	0.114	<0.008	<1.25E-06	<0.001[1]	<1.56E-07	<1.40E-6	>99.9988%
RSR	2,000	0.041	<0.008	<1.25E-06	<0.001[1]	<1.56E-07	<1.40E-6	>99.997%
RSRSSR	2,000	0.041	<0.008	<1.25E-06	<0.001[1]	<1.56E-07	<1.40E-6	>99.997%
EHHA	800	0.016	<0.008	<1.25E-06	<0.001[1]	<1.56E-07	<1.40E-6	>99.991%
EMPA	139,900	2.837	<20	<3.12E-03	<0.001[1]	<1.56E-07	<3.12E-3	99.89%
MPA	124,000[2]	2.515	53	8.27E-03	<0.001[1]	<1.56E-07	8.27E-3	99.7%
TOC	170,000	3.448	7	1.09E-03	<0.100[1]	<1.56E-05	1.11E-3	99.97%

(1) Estimate of detection limit obtained from SwR1.
(2) Assumes all organophosphate compounds are converted to MPA en route to full mineralization.

TABLE 8
Liquid Effluent Metals Analysis Results
for 8-Hour VX/NaOH Hydrolysate Test (from Ref. 3)

Sample No.	As (mg/l)	Ba (mg/l)	Ca (mg/1)	Cr (mg/l)	Cu (mg/l)	Fe (mg/l)	Hg (mg/l)	Mg (mg/l)	Ph (mg/l)	Se (mg/l)	Ti (mg/l)	Zn (mg/l)
Feed	0.125	0.236	121	1.38[1]	1.53	2.97[1]	0.004	2.79	0.50	2.0	0.25[1]	0.25
37[2]	---	---	---	0.15	---	0.2	---	---	---	---	2.4	---
61[2]	---	---	---	0.13	---	0.2	---	---	---	---	1.7	---

(1) After accounting for dilution of the feed (by a factor of about 8.2), the inlet and outlet concentrations of Cr and Fe are essentially equal, as expected. The Ti concentration in the effluent is significantly greater than that in the feed indicating some degree of corrosion, which was also expected given the Ti is not the recommended linear material.
(2) Liquid effluent samples 37 and 61 were analyzed for the metals listed.

at the reactor outlet at approximately room temperature]. It may be that it was not considered in the liquid analysis.

The metals - content of the SCWO product has not been examined extensively, Table 8; it should be the same as feed, with possible corrosion products added. The actual form of the metals has not been determined, (i.e. degree of oxidation, compound form).

The liquid product from the SCWO reactor is an aqueous solution of sodium sulfate and phosphate, and possibly small excess of sodium hydroxide. The salts will be recovered by an evaporator and filter, and the water recovered after evaporation. Any small residues of incompletely oxidized materials, e.g. unconverted MPA, will show up in these streams, - low volatility material in the salts, volatile materials in the water evaporated from the salts.

SCWO oxidation has been tested in a short pilot plant operation; the evaporator has been tested at a bench scale. The final plant operation has not been built or tested. An estimate of the compositions of the final products is shown in Table 9.

TABLE 9
VX PLANT EFFLUENTS (est'd)

Material	SCWO Vapor
H_2O	2.25%
CO_2	57.8%
N_2O	2.6%
O_2	37.3%
NO_x	7 ppm
CO	2.8 ppm
SO_2	21 ppm
VOC	0.16 ppm

	Solids from Evaporator	Water from Evaporator
Water	30.1%	99.96%
EMPA	181 ppm	.025 ppm
MPA	87 ppm	.012 ppm
Thiols	7 ppm	.001 ppm
TOC	73 ppm	.01 ppm
NO_3	450 ppm	0.061 ppm
HCO_3	6161 ppm	0.84 ppm
Na_2SO_4	34.6 %	47 ppm
Na_2HPO_4	34.6 %	47 ppm
NaH_2PO_4	0.00 %	0 ppm
H_3PO_4	0.00 %	0 ppm

The toxicity of the SCWO liquid product has been checked with animal tests, similar to those done on the hydrolysate, Table 10.

There were no signs of nerve agent action; the toxicity shown is extremely low and is simply due to the large salt content of the liquid from the SCWO process.

TABLE 10
Mouse Intravenous LD50/Toxicity Data VX - SCWO Product

Sample	pH	LD/50 (mg/kg) 24 hour	Toxic Signs
Clear Upper Layer	5.5	>10306	Ataxia, Decreased Activity
Homogenate	5.5	>10202	Ataxia, Decreased Activity

One of the remaining problems with the SCWO process is the process control required to ensure continuous high conversion level. As with baseline incineration, the first level of control is obtained by close control of operating variables: temperature, residence time (through put), excess oxygen level, and carbon monoxide level. The limiting material for conversion is the methyl phosphonic acid (MPA), and no real-time analytical procedure has been developed for monitoring it. (Unlike the base - line process where agent is the limiting material and is monitored continuously in close to real- time.) The MPA is soluble in water and is not highly volatile; most of any residual material should show up in the liquid from the SCWO process. This can be held long enough for analysis before going to evaporation and discharge. Any MPA in the vapor (part of the VOC shown in Table 9, 0.16 ppm) will probably be retained from discharge by carbon bed adsorption.

In summary: The combination of caustic neutralization of VX followed by supercritical water oxidation eliminates the extreme toxicity of VX and appears to yield products, which would be acceptable for discharge. The analytical results have not been carried out to the very low levels reported for base line technology, but are adequate for public safety.

5. Mustard Neutralization and Bio-Reaction

As with VX, it has been proposed to detoxify mustard by a neutralization process, in this case with hot water; caustic is added after the hydrolysis is complete. Mustard is insoluble in water, and vigorous mixing is required. The subsequent treatment required for the organic residue from hydrolysis would be a bio-reaction. The mustard contains several percent of chlorinated hydrocarbons, which are regulated materials and are resistant to bio-reaction; they will have to be separated from the bioreactor feed and treated separately. [Oxidation using UV-light and hydrogen peroxide has been proposed for them, but no final decision has been made.]

The bio-treatment proposed is a multi-stage treatment: first, hydrolyzed mustard together with some added nutrients is bio-treated in an aerobic digester; then the product mixed with other waste materials and treated again in an existing anaerobic bio treatment plant, (a "trickling filter").

The effluent materials of concern are: gas streams vented from the neutralization reactor and from the bio-reactors; solids from the first bio-reactor, a liquid stream from the final "trickling filter" of the existing bio-treatment plant; and the product resulting from the volatile chlorinated species separated from the mustard. In addition, the analysis of the neutralized material is of interest, to judge the effectiveness of the neutralization process in detoxifying the agent.

The primary hydrolysis reaction replaces the 'Cl' of the mustard to form a glycol (thiodiglycol - TDG):

$$S(CH_2\,CH_2\,Cl)_2 + 2H_20 = S\,(CH_2\,CH_2\,OH)_2 + 2HCl. \qquad (1$$

Side reactions can occur, involving the TDG for example, to produce complex sulfonium ion species. The mustard agent itself is impure (as low as 80% mustard), so that impurities will also contribute to the complexity of the neutralized mix. The presence of some sulfide or thiol compounds can result in an objectionable odor; addition of a little oxidizer (sodium hypochlorite) reduces this odor problem.

A typical analysis of liquid mustard is shown in Table 11, (based on GC/MS analysis). While there are probably many more components present at lower concentration (e.g. ppm), this analysis appears adequate to guide the following treatment.

TABLE 11
Table 3-14 Compounds in HD Agent

Compound	Area %
HD	91.38
Q*	6.08
2-chloroethyl 4-chlorobutyl sulfide	0.86
1, 4-dithiane	0.81
1, 2-dichloroethane	0.35
Bis 3-chloropropyl sulfide	0.18
2-chloropropyl 3'-chloropropyl sulfide	0.18
2-chloroethyl 3-chloropropyl sulfide	0.14
1-chloropropyl 2-chloroethyl sulfide	0.02
1, 4-thioxane	<0.01

Q* : Cyclic sulfonium ion

$CH_2 - CH_2$

$S \qquad S^+ - CH_2CH_2Cl$

$CH_2 - CH_2$

The mustard also contains small concentrations of metals, Table 12. The exact form is not known.

There is generally a residual 'heel' in mustard containers, a gel, which will not flow but can be washed out. It can amount to more than 10% of the stored agent. This material is also hydrolyzed and bio-treated. Analyses of three 'heel' samples are shown in Table 13. The 'heel' also contained metals, with a large amount of iron (e.g. >100000 ppm) as iron sulfide. The hot water clean-out material has been shown to have a mild deleterious effect on the aerobic bioreactor operation; COD-removal will be reduced a few percent - e.g. from 90 to 85% removal. No effect on the final trickling filter operation has been observed.

TABLE 12
Concentration of Metals in HD Agent

Metal	Content (ppm)
Aluminum	<13
Antimony	<9
Arsenic	18.9
Barium	<0.3
Beryllium	<0.2
Bismuth	<12
Cadmium	<3
Calcium	9.6
Chromium	<2
Cobalt	<2
Copper	91.8
Iron	5,035
Lead	<7
Magnesium	<0.4
Manganese	0.6
Nickel	<4
Phosphorus	32.6
Selenium	<14
Silicon	110.9
Silver	<4
Sodium	<16
Sulfur	264,420
Thallium	<13
Thorium	<11
Tin	<6
Vanadium	<3
Zinc	<0.7

A small amount of vapor is produced during the clean out of the mustard containers. An analysis of the air in the ton-container during clean-out shows some volatile organics (VOCs) as well as mustard, at the ppm level, Table 14. The actual quantity leaving the clean-out process is not known, but is small. It goes to a carbon adsorption bed. [Carbon is routinely used for cleaning the air flowing through a chem-weapons disposal plant.]

TABLE 13
Complex Organic Compounds in HD Heel by NMR (mole ratio)

Sample #	Location	HD	Q Cyclic Sulfonium Ion*	1, 4 Dithiane	Other
C-03-04-HH-1936	Top of peak	53	42	-	5
C-03-05-HH-1936	Bottom of heel	33	60	1	6
C-03-06-HH-1936	Middle portion	14	86	-	-

* where the Q cyclic sulfonium ion is as follows:

CH_2 - CH_2
S S^+ - CH_2CH_2Cl
CH_2 - CH_2

TABLE 14
Content of HD Agent and VOC's in Initial TC Vapor

Compound	Concentration (mg/L)
1, 2-dichloroethane	6.00
Bis-(2-chloroethyl) sulfide (HD)	0.83
1, 4-dithiane	0.6
2-Chlorobutane	0.21
Tetrachloroethene	0.083

6. HD Neutralization

The hydrolysis with hot water, followed by addition of sodium hydroxide, reduces the mustard to non-detect level, less than 20 ppb. Most of the mustard (99%) is converted to the thiodiglycol as expected. The organic compounds in the hydrolysate at different periods of the reaction, - after agent addition to the hot water, and after subsequent addition of NaOH, - are shown in Table 15. Only the major components were analyzed, since this hydrolysate is not a final product.

TABLE 15
Organic Compounds by ^1H NMR (Mole Percent)

Sample Time	(79 minutes after end of HD feed)	(63 minutes after end of NaOH addition)
TDG	90.7	91.2
CHTG	1.7	(0.05)
Q-OH	2.5	3.4
1, 4-dithiane	0.5	0.8
glycol	0.5	0.5
$C_6H_{12}SO_2$ compounds	2.1	2.2
acetone	(0.02)	---
other	2.0	1.8

CHTG = $HOCH_2CH_2$-S-CH_2CH_2-S^+-$(CH_2CH_2OH)_2$
Q-OH = $HOCH_2CH_2$-S-CH_2CH_2-S-CH_2CH_2OH

The analysis of Table 15 does not show the chlorinated materials that are not affected; they amount to a few percent of the original feed, e.g. 3%.

Most of the lighter components are stripped from the hydrolysate before the bioreactor. The nature of the materials and the fraction removed by the stripper are indicated in Table 16. No final decision has been made for handling these materials. Some disposal method will be required since they represent a substantial amount, a few percent of the original mustard.

The hydrolysate will be treated by aerobic biodigestion to produce a solid product and a liquid effluent. The solid product, the bio residue with some salts and metals, will be filtered, dried, and sent to a landfill. No detailed analytical data on the solid are available at this time. It will be treated as a toxic waste.

TABLE 16
Other Organics in Hydrolysate

Category	Compound	Formula
Nonbiodegradable VOCs: % Stripped from feed HD: 85%	1, 2-dichloroethane	$C_2H_4Cl_2$
	trichloroethylene	C_2HCl_3
	tetrachloroethylene	C_2Cl_4
	1, 1, 2, 2-tetrachloroethane	$C_2H_2Cl_4$
Biodegradable VOCs: % Stripped from feed HD = 5.0%	hexachloroethane	C_2Cl_6
	thiodiglycol	$C_4H_{10}O_2S$
Other Organics: % Stripped from feed HD = 0.4%	Cyclic sulfur compounds	
	Chlorosulfides	

Experimental work on a bioprocess for the hydrolysate has given encouraging results, with most regulated species in the liquid below the State Permit Limits - Table 17. The values for BOD, total N, and total P were above Permit limits however, so the material will be treated further in an existing anaerobic "trickling filter", part of the base sewage treatment plant.

TABLE 17
Comparison of average results for the SBR effluent during the test phase and the daily maximum discharge limits.

Parameter	SBR Effluent (mg/L)	Permit Limits (mg/L)
BOD_5	53.6 ± 8.1	20 *
Total N	21.0 ± 4.4	11 *
Total P	19.54 ± 3.37	2 *
Total Cyanide	< 0.005	1.20 **
Copper	<0.165	3.38 **
Cadmium	<0.065	0.69 **
Lead	<0.065	0.69 **
Nickel	<0.055	3.98 **
Zinc	0.312	2.61 **
Chromium	<0.100	2.77 **
Silver	<0.035	0.20 **
Total Toxic Organics	BQL ***	2.13 **
Total Petroleum Hydrocarbons	1.10	100 **

* Monthly Average Values. Source: State of Maryland, Aberdeen Proving Grounds, Edgewood Area, WWTP, and Effective February 1, 1992.
** Daily Maximum Values. Source: APGR 200-41, July 30, 1994
*** All TTOs were below the method quantitation limit

The "Total Toxic Organics" shown in Table 17 were examined for a large number of specific regulated materials; as usual, these are grouped as "semi-volatile" and "volatile" target compounds, as well as another category listed as "GC" target compounds. A few of those examined are shown in Table 18. Practically all were below the detection limit (at the parts per billion level). The BOD-figure (Table 17) shows that organics are present at the parts per million level; the specific materials have not been identified. Toxicity tests on aquatic species (shown later), indicate some small level of toxicity remaining in the liquid at this point.

The liquid from the aerobic bioreactor will be treated further in combination with other biologically treated material from the base sewage disposal plant.

Tests to date indicate that the design case will produce a final discharge from the trickling filter with BOD and total Nitrogen below NPDES limits, Table 19. The "Total P" shown in Table 19 exceeds the allowable state limit; it is current practice however, to precipitate phosphate chemically before discharge, so that all the parameters listed will be in compliance.

The aerobic bioreactors all produce a gas stream of excess air, carbon dioxide from the digestion process, and other possible material. This gas will be carbon treated before release to the atmosphere. No analytical data are available.

TABLE 18
Bio Reactor Effluent
Target Compounds (Many Not Shown)

Semi Volatile	Compound	Measured mg/L ppb/weight	Detection Limit mg/L ppb (weight)
	1, 4 Dichlorobenzone	BQL *	50.0
	2, 4, 5 Trichlorophenol	BQL	50.0
	2, 4 Dinitrotoluene	BQL	50.0
	o, m and p Cresol	BQL	50.0
	Pyridine	BQL	50.0
	1, 2 Dichlorobenzene	BQL	10.3
	2, 4 Dichlorophenol	BQL	10.3
	2 Methylnapnthalene	BQL	10.3
	Anthracene	BQL	10.3
Volatile			
	1, 1, 1 Trichloroethane	BQL	5.0
	1, 1 Dichloroethane	BQL	5.0
	1, 2 Dichloropropane	BQL	5.0
	2 Butanone	BQL	10.0
	Acetone	BQL	10.0
	Benzene	BQL	5.0
	Carbondisulfide	BQL	5.0
	Vinyl Acetate	BQL	10.0
	Vinylchloride	BQL	10.0

* Below Quantitation Limit (i.e. Detection Limit)

TABLE 19
Results

Trickling filter characteristics

Parameter (mg/L)	Test Feed	Effluent	Control Feed	Effluent	NPDES limit
BOD	31 ± 17	13 ± 4	12 ± 5	2 ± 0	20
COD	135 ± 3	67 ± 16	19	17 ± 23	na
Total N	7.5 ± 3.2	5.9 + 3.5	5.1 ± 2.4	3.7 ± 2.7	11
Total P	9.0 ± 1.9	8.8 ± 2.1	6.2 ± 1.3	7.5 ± 1.7	2 *

* FOTW removes phosphorous by chemical precipitation prior to discharge
BOD = biochemical oxygen demand N = nitrogen
COD = chemical oxygen P = phosphorus

7. Toxicity Testing

The liquid effluent from the plant, as well as some of the liquid streams within the plant, has been characterized for toxicity to some aquatic species. Both bioassays as well as chronic bioassays have been carried out on various species as follows:

BioAssays

MICROTOX Assay:	Vibrio fischeri
Daphnia Assay:	Daphnia magna
Shrimp Assay:	Artemia salina
Fish Assay:	Sheepshead minnow
	Fathead Minnow
Treatability Study:	FOTW Bio-mass

Chronic Bioassays

- Ceriodaphnia Survival and Reproduction Test
- Fathead Minnow Survival and Growth Test
- Sheepshead Minnow Survival and Growth Test

The tests show the hydrolyzed mustard solution to be toxic to some species, rather non-toxic to others. For example, Fathead minnows (a fresh-water species) showed an EC50 (50% death) of 12% vol./vol.; Sheepshead minnows (a salt water specie) had a much higher tolerance, 80% vol./vol.

The liquid product from the trickle filter (the material for final release), demonstrated extremely low toxicity, practically that of water. Tests for chronic toxicity, as measured by survival and reproduction rates, gave the results of Table 20.

TABLE 20
Effects of Bio Effluent on Ceriodaphnia

Survival (% Vol/Vol)

1st Aerobic Reactor		Trickle Filter Effluent	
NOEC *	LOEC *	NOEC	LOEC
9%	12%	100%	N/O

Reproduction (% Vol/Vol)			
3%	6%	50%	100%

* NOEC No Observable Effects
LOEC Lowest Observable Effects

8. Summary

The 'complete' mustard neutralization process has not yet been demonstrated. It has been shown that the mustard can be hydrolyzed almost quantitatively to the corresponding glycol by hot water and added caustic; the mustard is reduced to below detection level of 20 ppb. Subsequent treatment (an aerobic bio reactor followed by an anaerobic trickling filter) has yielded a liquid product suitable for discharge. Solid and gaseous products from the aerobic digester have not been examined in detail.

The process is complicated by the impurity of the original mustard. A large 'heel' of polymeric gel, rust etc.; left in the original storage tank after draining it must be cleaned out with a hot water spray. It has a mildly deleterious effect in the bio-reactor. In addition, the mustard contains some light chlorocarbons that are non-reactive, both to hydrolysis and to bio-degradation.

RISK ASSESSMENTS OF THE POTENTIAL HAZARD CONNECTED WITH THE OBJECTS OF STORAGE OF WARFARE CHEMICAL AGENTS

Kambarka, Kizner, and Shchuchye Arsenals

DR. VLADIMIR M. KOLODKIN
INTD, Udmurt State University
1 Universitetskaya St.,
Room 200/3, Build. IV
426034 Izhevsk, RUSSIA

1. Introduction

In the complex of problems linked to the storage and planned destruction of chemical weapons, to us, the inhabitants of the areas of storage and the future destruction of chemical weapons, the main problem is the problem of safety to the man and the environment. This problem cannot be solved without a preliminary analysis of all possible aspects of the potential damage. In this connection any decision made (or not) relative to the object with chemical weapons implies a preliminary analysis of the potential hazard conditioned by the decision.

One may suppose that the level of hazard linked to the objects of storage, on condition of operation without accidents, is insignificant. This is confirmed by the fact that there are no grounds to state that the operation of the objects of storage influenced the state of health of the population or the state of the environment. This statement relative to the objects of storage situated on the territory of the Udmurt Republic is based on the results of check-ups made by various organizations independently [1-2] and the results of spot checks of the state of the environment [3]. Those results were obtained despite the fact that the objects of storage situated on the territory of the Udmurt Republic have been operating for several tens of years.

The accident hazard engendered by release is more significant. The causes leading to an accident at the object with chemical agents may be various. This can be such a little probable event like a fall of a meteorite or an air crash leading to a fall of the plane on the territory of the arsenal. It can result from an act of terrorism. It can be a disaster of the technogenic or natural origin. It is not possible to exclude from consideration such a factor like errors by personnel during maintenance works at the object.

The general diagram of the potential accident hazard linked to the objects of storage of warfare chemical agents is shown in Fig. 1. According to the diagram, the analysis includes:

- a determination of the conditions under which the source of the toxic hazard appears and a determination of the characteristics of the source;
- a prognosis of time-space characteristics of contamination of the abiotic

F.W. Holm (ed.), Effluents from Alternative Demilitarization Technologies, 121–139.

medium;

- a prognosis of the frequency of the origin of accident situations;
- and a calculation of risk assessments as characteristics of the potential hazard.

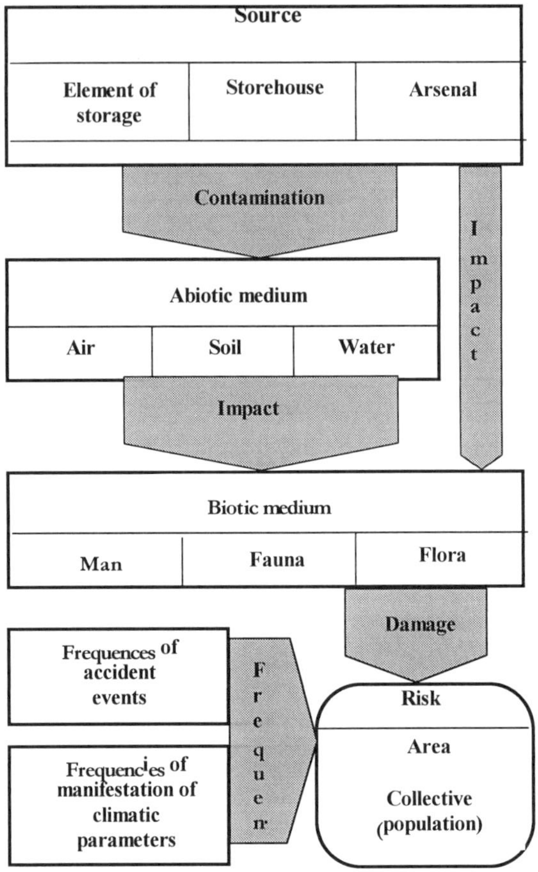

Figure1. General diagram of the prognosis of the level of the potential accident hazard

2. Characteristics of the accident hazard

2.1. ACCIDENT RISK ASSESSMENTS

Risk assessments are the quantitative characteristic of the level of the potential hazard. The variety of manifestation of the accident hazard is matched by a variety of assessments. The level of hazard manifested at a certain point on the territory may be described by the magnitudes of the area and collective risks. By its meaning the area risk assessment corresponds to the predicted frequency of affection of the recipient of risk at a certain point [4]. The collective (population) risk characterizes the level of the potential hazard considering the number of recipients of risk relative to the point. To its meaning the magnitude of the collective risk corresponds to the number of recipients affected within a time interval on a certain territory. The integral collective accident risk assessment corresponds to the number of recipients of risk affected on a certain territory.

The determining relation for the prognosis of risk assessments [5] can be represented in the following form:

$$\boxed{\text{Accident risk assessments}} = \sum_{s} \boxed{\text{Frequency of s-th accident process}} \times \boxed{\text{Damage (losses) with s-th accident process}} \quad (1)$$

It follows from the above relation that the prognosis of the level of the accident hazard has two aspects: one aspect is linked to the frequency analysis of accident processes, and another to the prognosis of damage conditioned by accident processes. The accident process corresponds to a certain accident scenario. It can be represented as a trajectory in the space of events. Every trajectory includes an initiating event (incident), a sequence of post-incident events, the final event (realization of the incident). By a final event is meant an elementary accident event (a spill, an explosion, a fire) or a sum of them. The events are characterized by frequencies (probabilities), which, in the general case, depend on the situation of the event on the trajectory. Let us introduce the following designations: F_s is the frequency of the s-th initiating event; P_i^s is a probability of development of the accident scenario along the trajectory i connecting the incident s with the realization of the incident.

According to the designations introduced the expression for risk assessment is represented in the view:

$$\boxed{\text{Accident risk assessment}} = \sum_{s} \cdot \sum_{i} F_s \cdot P_i^s \cdot W^{s,i} \quad , \quad (2)$$

where $W^{s,i} = \sum_{k} \nu_k \cdot U_k^{s,i}$ is a conditional accident risk assessment. $W^{s,i}$ corresponds to the condition of development of the accident situation from the initiating event s along the trajectory i. ν_k is a frequency of the state of the environment characterized by the k-th set of climatic parameters. $U_k^{s,i}$ is a magnitude of damage of the final event on condition that the accident situation developed from the initiating event s along the trajectory i with the k-th set of climatic parameters. The magnitude of the conditional risk in every specific case has its own expression.

Let us introduce a polar coordinate system (r,φ) with the origin at the point of the centre of the source of the toxic hazard.In the special case with the toxic action on the man by means of the atmospheric air in case of realization of a specific accident scenario for the conditional area accident risk assessment $W^{s,i}(r, \varphi)$ we have the following:

$$= \int_0^{2\pi} d\alpha \int_0^{t_a} dt \int_0^{V_{max}} dV \int_{T_{min}}^{T_{max}} \mu(\alpha)\lambda(V)\cdot(\sum_i P(V \in \delta V \mid d_i))\,\bar{\eta}(t)\,\chi(T) \times$$

$$\times U(r,\varphi,d_i,t,\alpha,V,T)\,dT\,, \tag{3}$$

where $\mu(\alpha)$ is the density of distribution of probabilities of wind directions depending on the magnitude of the value of the angle α; $\lambda(V)$ the density of distribution of the probabilities of values of the wind velocity V; $P(V\in\delta V|d_i)$ the probability of the fact that with the wind velocity $V\in\delta V$ the state of the atmospheric stability belongs to the stability class d_i; $\bar{\eta}(t)$ the density of distribution of the probabilities of the absence of precipitation in the form of rain; t_a the temporary scale of the toxic action with the transmission of the action by means of the atmospheric air; $\chi(T)$ the density of distribution of the probabilities of values of the temperature of the atmospheric air; $U(r,\varphi,d_i,t,\alpha,V,T)$ the function characterizing damage during the realization of a specific accident scenario.

The conditional collective accident risk assessment at a point $(r,\varphi) \in \Omega$

$$W_g^{s,i}(r,\varphi) = \psi(r,\varphi)\,W^{s,i}(r,\varphi). \tag{4}$$

where $\psi(r,\varphi)$ is the density of distribution of people in the zone of affection Ω, $(r,\varphi) \in \Omega$, is the total of people in the area Ω. The integral magnitude of the conditional collective accident risk assessment over the zone of affection $W_G^{s,i}(\Omega)$:

$$W_G^{s,i}(\Omega) = \int_\Omega W_g^{s,i}(r,\varphi)\,d\Omega\,. \tag{5}$$

2.2. DAMAGE FROM AN ACCIDENT

We shall characterize damage from an accident by a relative share of the population affected. Let us assume that affection is conditioned by the action of one and the most toxic agent, and only one way of transmission of the toxic action to the recipient of risk is important.

In connection with organophosphorus agents the dependence "dose-effect" may be approximated by the function [6] :

$$U(d) = 1/[1+(d/LC_{\tau 50})^{\beta}], \tag{6}$$

where U(d) is the quantitative assessment of the effect of the dose d; $LC_{\tau 50}$ the toxic dose corresponding to a 50% level of the lethal affection; β = 1.677 / Ln(S); S the function of toxicity. With the inhalation action we have the expression $d = C^n \cdot t$ for the toxic dose d[7]. For organophosphorus agents the index of the power n was taken equal to 1.

TABLE 1
Numerical values of the toxic dose $LC_{\tau 50}$, mg*min / m^3

	Sarin	Soman	V-gas	Lewisite
Literature data	70-100[8]	40-70[8]	35-45[8]; 10[9]; 35-36[10];	1'200[11]; 1'500[11]; 1'300[12]
Used for calculation	100	50	35	1200

The function of toxicity S characterizing the resistance of the organism to the toxic action was taken in the following view

$$S = 0.5\ (LD_{84}/LD_{50} + LD_{50}/LD_{16}), \tag{7}$$

where LD_{84}, LD_{16} are the average doses producing the lethal affection of 84% or 16% correspondingly. For the organophosphorus agents the values of the function S are in the range from 1.1 to 4.4.

Table 1 shows the values $LC\tau_{50}$ from literature and the values used in calculations. A probit-model was used for phosgene and Lewisite[12]

$$Pr = A + B*Ln(C^n t), \tag{8}$$

where for phosgene by the inhalation action[12]: A = -19.27; B = 3.686; n = 1.The values used for Lewisite [11]: A = - 3.29, B = 1.0, n = 1.25.

Expression (8) gives a relation of the toxic dose with the probit-variable Pr. The percent of people affected is determined by using the standard probit-tables [7].

2.3. CHARACTERISTICS USED WHEN COMPARING THE OBJECTS OF STORAGE TO THE LEVEL OF THE POTENTIAL HAZARD ENGENDERED BY THE OBJECTS

To compare the objects of storage of warfare chemical agents to the level of the potential accident hazard the following characteristics are used.

Integral magnitude of the collective accident risk R_G.

$$R_G = \sum_s \cdot \sum_i F_s \cdot P_i^s \cdot W_G^{s,i} \tag{9}$$

The integral conditional collective risk assessment $W_G^{s,i}$ corresponds to the expression (5). To its meaning the characteristic (9) corresponds to the probable assessment of the number of victims linked to a potentially hazardous object. If the frequency characteristics correspond to a time-interval of one year, the probable assessment of the number of victims also corresponds to an annual interval.

Predicted frequency of affection of at least one person P^{let}. It is supposed that a person may be at any point of the area of affection Ω. P^{let} is calculated relative to the point with a maximal level of hazard (It is a point of the centre of the source of the toxic hazard). The measure of affection is a fatality. The level of affection is limited by the magnitude m (in this case it is supposed that m = 50 %).

$$P^{let} = \sum_s \cdot \sum_i F_s \cdot P_i^s \cdot \eta_w^{s,i} \tag{10}$$

where $\eta_w^{s,i}(r,\varphi) =$

$$= \int_0^{2\pi} d\alpha \int_0^{t_a} dt \int_0^{V_{max}} dV \int_{T_{min}}^{T_{max}} \mu(\alpha)\lambda(V) \cdot (\sum_i P(V \in \delta V \,|d_i))\, \bar{\eta}(t)\, \chi(T) \times \sigma^{let}$$

$$\sigma^{let} = \begin{cases} 1, \text{ if } & d(r,\varphi,d_i,t,\alpha,V,T) \geq d_m,\ (r,\varphi) \in \Omega \\ 0, \text{ if } & d(r,\varphi,d_i,t,\alpha,V,T) < d_m,\ (r,\varphi) \in \Omega \end{cases}$$

$d(r,\varphi,d_i,t,\alpha,V,T)$ is a toxic dose to which a recipient of risk was exposed by a specific scenario, d_m the toxic dose corresponding to the level of affection m ($d_m = d_{50}$).

When forecasting the frequency of affection of at least one person in a populated area P_p^{let}, the analysis is limited by the zone of the populated area Ωp : $(r,\varphi) \in$ Ωp.

*Parameters of the zone of affection corresponding to the level of risk R**. The zone of affection Ω is characterized by the area $S(R^*)$ and the effective radius $R_{eff}(R^*)$.

The field of values of the area accident risk in the area Ω according (2) is

$$R(r,\varphi) = \sum_s \cdot \sum_i F_s \cdot P_i^s \cdot W^{s,i}(r,\varphi) \tag{11}$$

where $W^{s,i}(r,\varphi)$ corresponds to the expression (3). The zone of affection corresponding to the level of risk R^* is marked off by the condition $R(r,\varphi) > R^*$. The area of the zone is determined in accordance with the topographic base of the area of situation of the object. If several levels of risk are considered, the area of affection is differentiated to the level of the potential hazard.

Characteristics of the maximal affecting effect by an accident. For a specific accident scenario the following is forecasted:

- the maximal number of fatalities N_{let}

$$N_{let} = \max \int_\Omega M \cdot \psi(r,\varphi) \cdot U(r,\varphi,d_i,t,\alpha,V,T)\, d\Omega \ , \tag{12}$$

where M is a number of people in the area Ω, ψ(r,φ) the density of distribution of people over the area Ω, $U(r,\varphi,d_i,t,\alpha,V,T)$ the function of damage by an accident. Climatic parameters correspond to a maximal affecting effect. The level of repetitiveness of climatic parameters is limited below by the magnitude k_{low}.

- the maximal distance at which the dose corresponding to the affection of the n-th part of the population is achieved, L_n^{max}.

The value L_n^{max} follows from the solution of the equation

$$\max U(L_n^{max}, \varphi, d_i, t, \alpha, V, T) = n, \tag{13}$$

where the set of climatic parameters corresponds to a maximal affection. The level of repetitiveness of climatic parameters is not lower than k_{low}.

3. Sources of the accident toxic hazard. Conditions of their origin. Assessments of parameters

The origin of the source of hazard, the transmission of the toxic action to the recipient of risk, the reaction of the recipient of risk to the toxic action are described by the corresponding mathematical models [13-14]. The models are concretized depending on the type of toxic agents, the conditions of storage, the conditions of dispersion, an so on. In particular, in Fig. 2 there is shown a system of mathematical models oriented to the analysis of the level of the accident hazard as applied to the conditions of Kambarka, Kizner, and Shchuchye Arsenals.

The models take into account the most important factors. For example, the model of depressurization of ammunition with organophosphorus agents by a fire takes into account that on heating organophosphorus compounds are decomposed with release of gas products. The release of gas products leads to an increase in the pressure of the inner space of the shell. When the pressure achieves a certain magnitude the integrity of the material of the inner (ignition) cup cannot be conserved anymore, and the destruction products and not decomposed substances may be ejected into the environment. If the temperature of the environment is lower than the temperature of ignition of destruction gas products, a source of toxic hazard is formed.

The parameters of the sources of the toxic hazard were predicted by three elementary accident events - a spill of toxic agents, an explosion at the object of storage, a fire.

3.1. SPILL OF TOXIC AGENTS

The source of the toxic hazard, in the case of a spill, is characterized by the area of the spill and the intensity of evaporation.

The intensity of evaporation is determined by the characteristics of molecular processes at the boundary of phase separation and the processes of transfer in the atmosphere. These processes are interconnected and are modeled within a common mathematical model. The model of inter-phase processes corresponds to the molecular kinetic theory of evaporation. The transport of admixture in the atmosphere is described by the equation of turbulent diffusion [13-14].

The input parameters of the model of evaporation under consideration are the individual physical properties of the evaporating substance (the coefficient of molecular diffusion, the enthalpy of evaporation, the pressure of saturated vapor); the atmospheric state (the wind velocity, the class of stability); dimensions of the area of spill; temperature.

	**Models of accident events**	- Model of fire in storehouse - Model of forest fire - Model of explosion - Model of evaporation - Model of burning of liquid
	**Models of origin of source of hazard**	- Model of depressurization of shell by fire - Model of depressurization of shell by explosion - Model of destruction of container with liquid by explosion
	**Abiotic medium**	- Model of turbulent diffusion in atmosphere - Model of filtration in soil - Model of transport of admixture by river flow - Model of spreading of liquid
	**Biotic medium**	Models of reaction of recipients of risk (man, fauna, flora) to the toxic action
!	**Level of potential hazard**	Mathematical representation of accident risk assessments relative to: man, fauna, flora

Figure 2. Complex of mathematical models used for building accident risk assessments.

The results of modeling show that the average to the spill area intensity of evaporation: a) slightly changes when the linear dimensions of the spill surface change; b) changes considerably when the atmospheric state changes; c) depends on temperature. With all other coinciding conditions the intensity of evaporation is proportional to the pressure of saturated vapor which depends on temperature.

As an example Table 2 shows the results of calculation of the average to surface intensity of evaporation of Sarin for a square spill area of 1'000 m^2.

TABLE 2
Intensity of evaporation of Sarin $g/(m^2\ s)$
with a wind velocity V, temperature T and various states of the atmospheric stability

V, [m/s]	Class of the atmospheric stability							T, [^{0}]
	1	2	3	4	5	6	7	
1	0.019 0.092	0.016 0.074	0.013 0.060	0.009 0.043	0.008 0.036	0.006 0.027	0.005 0.023	0 20
2	0.034 0.16	0.028 0.13	0.023 0.11	0.017 0.079	0.014 0.066	0.010 0.050	0.009 0.043	0 20
3	0.046 0.22	0.038 0.18	0.031 0.15	0.023 0.11	0.019 0.093	0.015 0.070		0 20
4		0.047 0.22	0.039 0.19	0.029 0.14	0.025 0.12			0 20
5			0.046 0.22	0.035 0.17				0 20
6			0.052 0.25	0.040 0.19				0 20
8			0.063 0.30	0.049 0.23				0 20
10			0.072 0.35	0.057 0.27				0 20
12			0.080 0.38	0.064 0.31				0 20
16			0.092 0.44	0.076 0.36				0 20

The assessments of the average to the spill surface intensity of evaporation of various toxic agents are shown in Table 3.

TABLE 3
Intensity of evaporation, $[g/(m^2\ s)]$ (wind velocity 4 m/s, class of the atmospheric stability 4, temperature 20^0)

Lewisite	Sarin	Soman	V-gas
55	140	25	0.02

3.2. EXPLOSION AT AN OBJECT OF STORAGE OF CHEMICAL AMMUNITION WITH ORGANOPHOSPHORUS AGENTS

For the case of an explosion at an object of storage stress arising in the material of the inner cup of the chemical ammunition by the passage of the shock wave has been calculated. When the magnitude of stress achieved the yield point it was assumed that the toxic agents were released into the environment. To assess the consequences of the explosion of a charge with a certain power inside the storehouse Monte-Carlo procedure of computer modelling of the scattering of shells with depressurization was applied. It was taken into account that part of the chemical ammunition will be kept by not destructed walls of the storehouse. The positive pressure for the fall of walls at the front of the shock wave was taken 0.5 atm, [15]. The data on the positive pressure at the front of the shock wave, the speed and the density of the detonation product corresponds to [16].

TABLE 4
Number of depressurized shells N ejected to a distance r from the storehouse by an explosion of a charge with a mass m in trotyl equivalent

Mass EA, [kg]	Inside the storehouse	Distance r, [m]					
		50	100	150	200	300	400
10	180	4	2	1	0	0	0
50	230	20	9	6	3	1	1

TABLE 5
Characteristics of the zone of contamination by explosions at storehouses

Arsenal	Mass EA [kg]	Radius of the zone of contamination, [m]	Mass CA in the zone of contamination, [kg]		
			Sarin	Soman	Total
Kizner	5	50	103	36	139
Kizner	50	340	471	162	633
Kizner	100	620	921	316	1237
Shchuchye	5	50	93	61	154
Shchuchye	50	340	421	282	703
Shchuchye	100	620	826	549	1375

Some assessments of the quantity of depressurized shells ejected from Shchuchye Arsenal to a distance *r* and covering the area in the form of a strip 10 m large are given in Table 4. It was assumed that the shells of various calibers were located in the storehouse arbitrarily, and that is why the data of the table are given taking into account the caliber averaging.

Table 5 shows assessments of some characteristics of the zones of contamination. By a zone of contamination is meant a territory on which a shell falls and gets depressurized. It was assumed that when the shell falls down its chemical agent flows out of the inner space in full.

An explosion at a storehouse of Shchuchye Arsenal close to a container with a missile warhead filled with V-gas leads to a depressurization and further outflow of the liquid. The assessment of the area of the spill mirror is 130 m^2. It was assumed that the explosion of a shell with a trotyl mass M would lead to a dispersion of 50∗M V-gas in the atmosphere. 5% of the mass of the dispersed liquid will remain in the air in the form of an aerosol cloud, and the rest will fall down on to the ground. The dimensions of the cloud have been evaluated numerically as the characteristic dimensions of scattering of detonation products [16].

3.3. AN EXPLOSION AT AN OBJECT OF STORAGE OF LARGE CONTAINERS (KAMBARKA ARSENAL)

The prognosis of destruction of a tank with Lewisite by an explosion at Kambarka Arsenal is based on the experimental Δp - I diagram of destruction (Δp is a maximal positive pressure at the front of the shock wave; I the pulse of the compression phase) [17].

The assessment of the area of spill of liquid by destruction is approximated by the regressive dependence [13]: $S(V)=a \cdot V^b$, where V is the volume of the liquid spilt, [m^3]; S the surface, [m^2]. The values of the regression coefficients: a= 53, b= 0.63.

TABLE 6

Assessments of the parameters corresponding to an explosion at Kambarka Arsenal

Mass of EA in trotyl equivalent, [kg]	Spill area, [m^2]	Number of tanks destroyed, [pc]	Linear size of the aerosol cloud, [m]	Mass of CA in the aerosol cloud, [kg]
1	540	1	9	2.2
10	2'500	11	17	22
50	3'900	16	35	225

The size of the aerosol release from a container situated immediately over the epicentrum of the explosion was evaluated as the size of the cloud of detonation products [16]. The mass of aerosol particles elevated into the air by the explosion is proportional to the mass of the explosive agent with the coefficient of proportionality 50. In doing so it was assumed that 10 % of the energy of the explosion goes into the ground by an explosion on a concrete base [18]. Table 6 shows the consequences of an explosion at Kambarka Arsenal under prognosis.

3.4. A FIRE AT A STOREHOUSE WITH ORGANOPHOSPHORUS AGENTS

In modeling the fire at a storehouse with chemical ammunition an empirical model of a fire in industrial buildings developed in [19] was used. The model allows us to calculate the total time of fire and the average temperature of fire as a time function. The assessment of the time of fire gives a magnitude of 2.8 hours for a wooden building and 2.1 hours for a concrete (brick) building. The dynamics of destruction of the chemical agents in the shells and the magnitude of release of the toxic agents into the atmosphere was evaluated taking into account the dynamics of temperature in the burning storehouse [14,20].

TABLE 7

Time interval till depressurization of chemical ammunition t_{min} and the mass of the agent M released by a fire

Type CA, (caliber)	t_{min}, [min]	M, [kg]
Sarin (122 mm)	21	0.4
Sarin (152 mm)	27	1.0
Soman (122 mm)	44	0.2
Soman (152 mm)	58	0,8

TABLE 8

Mass of chemical agents released from chemical ammunition by a fire

Arsenal	Type of the storehouse	Mass of chemical agents released by a fire, [kg]		
		Sarin	Soman	Total
Kizner	Concrete	41	10	51
Kizner	Wooden	44	9	53
Shchuchye	Concrete	43	18	61
Shchuchye	Wooden	40	16	56

Some characteristics of the sources of toxic hazard by fire are shown in Table 7: t_{min} is the design time to the depressurization of a unit of chemical ammunition on condition of the location of a shell in the zone of the flame; M is the design mass of chemical agent ejected from the shell by its depressurization.

The data shown in Table 8 allows for a comparison of the magnitude of release of chemical agents (Sarin, Soman,) from chemical ammunition by a fire. It follows from the data that the type of the storehouse does not influence considerably the power of the source of the toxic hazard arising as a result of puncture of chemical ammunition in the storehouse during a fire. Let us notice that the probability of opening of a shell filled with V-gas is insignificant.

By a fire at a storehouse of Shchuchye Arsenal the missile warheads filled with V-gas are the most dangerous. The rate of destruction would be considerably lower than in other kinds of chemical ammunition because of its housing and the considerable mass of the chemical agent. On the other hand, the relatively thin walls and this considerable mass of chemical agent will contribute to depressurization. The assessments show that the time interval till depressurization consists about one hour in the conditions of a developed fire.

4. Assessments of the toxic hazard for a case of terrorism

Warfare chemical agents generate a specific kind of hazard linked to their illegal use.[21]. The level of hazard as a result of an illegal use may vary widely. The place of use may be far away from the object of storage. In this chapter there are given assessments of the level of possible consequences by an illegal use of chemical ammunition filled with Sarin.

4.1. CONSEQUENCES OF A RELEASE OF CHEMICAL AGENTS INTO A WATER-RESERVOIR

To contaminate a water-reservoir with the dimensions $R \sim 1 \div 10\ km$ and a depth of $h \sim 1 \div 10\,m$ (pond, lake) to a dangerous concentration it is necessary to have ~ 100-1'000 units of chemical ammunition (300-3000 kg Sarin).

4.2. CONSEQUENCES OF A RELEASE OF CHEMICAL AGENTS INTO A WATER SUPPLY SYSTEM

The assessment of people affected gives a magnitude about 100 people if the chemical agents from chemical ammunition are introduced into the water supply system of a town.

4.3. POSSIBILITY OF CREATION OF DANGEROUS CONCENTRATIONS IN ROOMS

One unit of chemical ammunition can create a dangerous concentration in a closed room (a metro station) with a volume of 10^4 - 10^6 m^3.

5. Dispersion of toxic substances in the abiotic medium

Mathematical models describing the dispersion of toxic agents in the abiotic medium reflect the laws of mass, energy and pulse conservation.

5.1. ASSESSMENTS OF THE DEPTH OF PENETRATION OF THE TOXIC AGENTS INTO THE SOIL.

Warfare chemical agents that reached the ground surface would penetrate into the soil through cracks, inter-aggregate pores, and also the pores formed by plant roots. Suction into the soil proceeds under the action of the capillary force and the gravity. With a small volume of liquid only moistening of the surface layer takes place.

The upper and the lower assessments of the depth of penetration follow from the balance ratios: the lower L_1 considering the hydrolysis reaction, the lower L_2 considering the soil porosity. Some results are given in Table 9. In calculation the values of porosity m_0 and humidity (field moisture capacity FMC) typical to the given kind of soils were taken into account.

TABLE 9
Assessments of the ratio of the depth of penetration of chemical agents L to the height of the layer of the spilt liquid H for various soils

CA type	Soddy-podzolic m_o=0.44 FMC=0.24		Chestnut m_0=0.43 FMC=0.24		Chernozem m_0=0.55 FMC=0.34	
	L_1/H	L_2/H	L_1/H	L_2/H	L_1/H	L_2/H
Sarin	1.12	18.69	1.15	19.12	0.65	17.22
Soman	0.88	18.69	0.90	19.12	0.50	17.22
V-gas	0.61	18.69	0.62	19.12	0.35	17.22

With a height of the spilt layer of 5 cm the depth of penetration of the toxic substances into the soil does not exceed 1 m.

5.2. ASSESSMENTS OF DISPERSION OF CHEMICAL AGENTS IN THE ATMOSPHERE

The process of dispersion of a chemical agent in the atmosphere is modeled based on the equation of turbulent diffusion which describes the change in the concentration of the chemical agent in the atmosphere with time due to the convective transport by the middle wind and the processes of turbulent stirring.

The state of the atmosphere is classified by the degree of stability. In the model considered the state of the atmosphere is characterized by vertical profiles of the average wind velocity and the coefficient of turbulent diffusion. These characteristics can be restored based on data of standard ground meteorological observations.

The input parameters of the model are individual characteristics of the source of the chemical agent and the atmospheric state. The time dependence of the distance L along the direction of the wind, where the toxic dose corresponding to the 50% level of affection is achieved, is shown in Fig. 3.

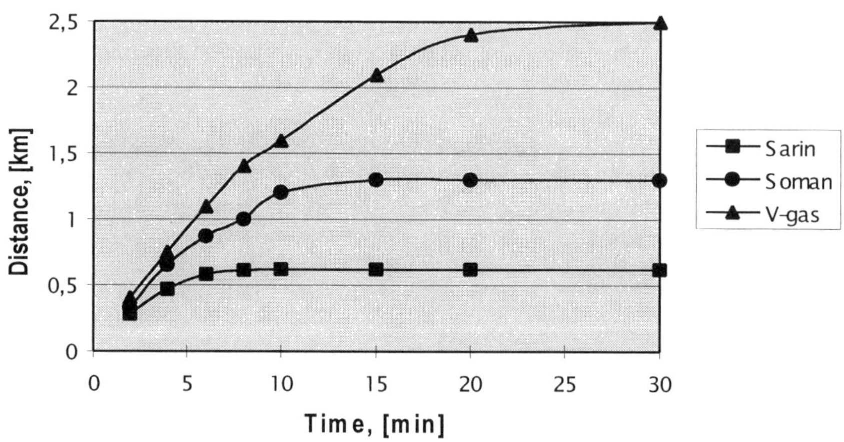

Figure 3. Distance L at which the toxic dose corresponding to the 50 % affection is achieved depending on the time interval Δt.

There is considered a situation of a single release of chemical agent into the atmosphere. It is assumed that the mass of the toxic substance is 60 kg; the volume of the cloud formed as a result of the release 1'000 m^3; the wind velocity 2 m/s; the class of the atmospheric stability 4. The time interval Δt is calculated from the instant of release.

6. Frequency characteristics of accident situations

The trajectory in the space of events corresponding to a certain scenario of an accident situation is characterized by two sets of frequencies: frequencies of events, frequencies of climatic factors (air temperature, humidity, wind direction, and so on).

Table 10 shows the frequencies under prognosis (probabilities) of some events that can lead to accidents at the objects of storage. The frequencies of events vary in a wide range. Based on the data of Table 10 the numerical assessment of the frequency of the origin of accident situations at the objects of storage of warfare chemical agents conditioned by natural disasters was taken 10^{-4} per year, i.e. one accident per 10'000 years. Let us stress that the assessment does not take into account the predicted frequency of illegal actions.

TABLE 10

Predicted frequencies of events initiating accidents

Initiating event	Predicted frequency, [$year^{-1}$]
Earthquake (Ural Region, Russia)	10^{-12} [22]
Storm (Ural Region, Russia)	10^{-4} [22]
Air crash (Kambarka Arsenal)	10^{-8} [22]
Fire in case of a spill of combustible reactants at chemical-technological objects	10^{-5} [23]

The frequency of manifestations of climatic factors follows from the analysis of the climatic characteristics of the chemical warfare agent storage sites.

7. Characteristics of the accident hazard engendered by the objects with warfare chemical agents

The mathematical models of the sources of the toxic hazard and the dispersion of the toxic substances in the environment are a basis of the computer system of assessment of the level of the potential hazard connected with the objects of storage of warfare chemical agent [24]. Some assessments of the level of the potential accident hazard are shown in Table 11. The characteristics correspond to Kambarka, Kizner, and Shchuchye Arsenals of storage of warfare chemical agents. For a preliminary analysis of the level of hazard connected with the destruction (processing) of chemical agent stockpiled at Shchuchye

Arsenal, the following assumptions were used:

- the supposed building site is situated at a distance of 11 km north of the object of storage;
- delivery access is a road which is 2 km distant from the populated areas.
- there are 100 people at the object of destruction at the instant of the accident.

TABLE 11
Assessments of the characteristics of the accident hazard connected with the objects of storage and destruction of warfare chemical agents

Object	$R_G(25)$, [people]	$P^{let}(25)$	$S^{let}(R>0.01)$, [km^2]	N_{let}, [people]
Kambarka	10^{-3}	10^{-3}	<1	<10
Kizner	10^{-2}	10^{-3}	5	700
Shchuchye (storage)	10^{-1}	10^{-3}	5	3000
	R_G, [people]	P^{let}	$S^{let}(R>0.01)$, [km^2]	N_{let}, [people]
Shchuchye (destruction)	10^{-2}	10^{-4}	2	300

The table contains the following designations: $R_G(25)$ is the integral magnitude of the collective risk per 25 years; $P^{let}(25)$ the predicted frequency of affection of at least one person during 25 years; $S^{let}(R>0.01)$ the area of the zone of affection corresponding to the level of the area risk $R > 0.01$; N_{let} the maximal number of fatalities under the less favorable meteorological conditions which frequency cannot be less than 0.1.

In calculating the magnitudes R_G the assessments of the characteristic frequencies of accident situations were used: 10^{-4} for the object of storage, $2.5 \bullet 10^{-3}$ for the object of destruction.

To assess the maximal affecting effects the following values of the climatic factors were used: the wind velocity of 8 m/s; the indifferent state of the atmosphere; the air temperature of 20^{0} ; the wind direction corresponding to the direction to the nearest largest populated area (Kambarka, Kizner, Shchuchye).

Fig. 4 shows the location of the levels of the area accident risk relative to Shchuchye Arsenal. The line of the level limits the territory at every point of which the value of the area risk $W(r,\varphi)$ is not less than the shown magnitude W^*. On the marked off territory the predicted frequency of affection of the man during a year is not less than the magnitude W^*.

Fig. 5 shows how the value of the integral collective accident risk assessment changes depending on the location of the potentially hazardous object. Every point of the area (r,φ) is matched by a value of the integral collective accident risk assessment, which can be interpreted as a prognosis of the frequency of affection of people if the potentially hazardous object is situated at a point (r,φ). By a potentially hazardous object is meant an object the level of the accident hazard of which is comparable with the level of the accident hazard of Shchuchye Arsenal. The given figure may be used, for example, in choosing the site for the chemical agent to be disposed of.

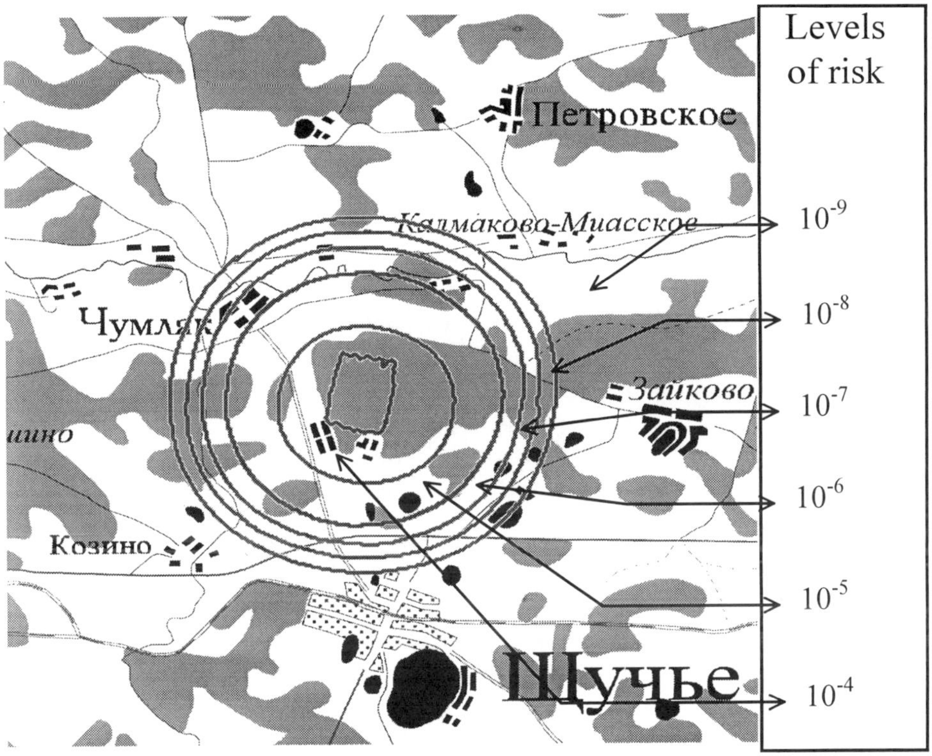

Figure 4. Levels of the area accident risk relative to Shchuchye Arsenal

8. Conclusions

The prognosis of accident consequences expressed in the categories of accident risk assessments determines the strategy of works on the safety and the environmental protection in the areas of storage and destruction of chemical weapons.

The results of the prognosis are a basis for:

1) The comparison of the importance of the objects of storage of warfare chemical weapons to the degree of the potential hazard relative to the man and the state of the environment. The differentiation of the territory of the areas of storage to the degree of the potential hazard.
2) The conduction of works on the determination of the zones of protective measures for the objects with warfare chemical agents.
3) The creation of information support for the solution of the problems of safety at the objects with warfare chemical agents, in particular:
   a) the creation of software for decision-making under the conditions of emergencies;
   b) the development of plans of action in the case of emergencies;

c) the creation of software to support training of the ministries and departments whose task it is to act in the condition of emergencies;
d) the conduction of works with the inhabitants of the stockpiling areas with the aim of decreasing the level of the social tension.

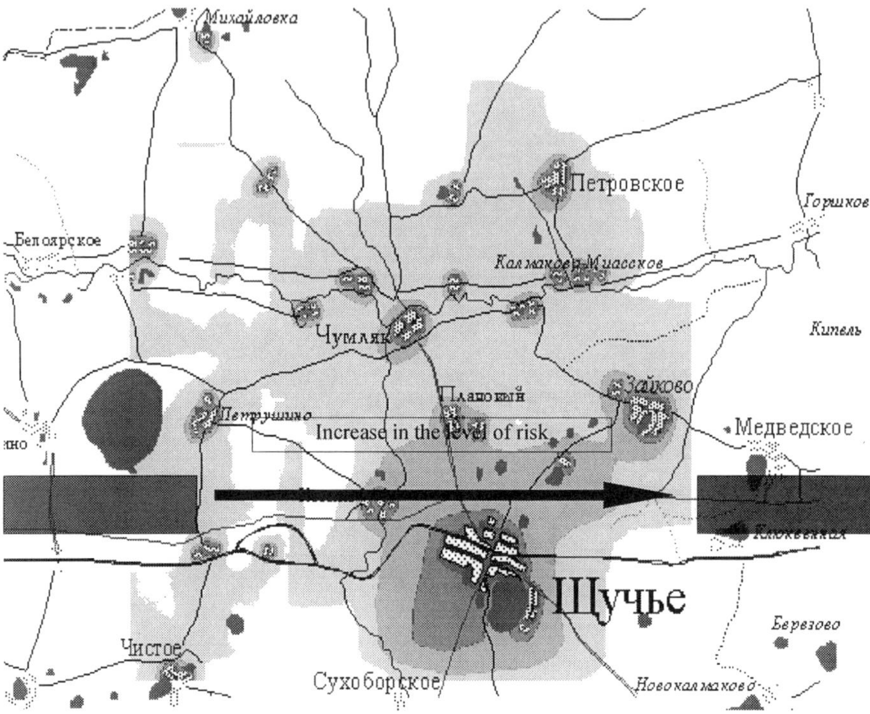

Figure 5. Dependence of the values of integral assessments of the collective accident risk on the accident site

9. References

1. *KiryanovN.A., Danilov G.E., Yezhova N.N, Kolesnikova M.B.* Evaluation of the state of health of the inhabitants of the Kambarka district resulted from the check-ups made by NIIGPT, VNCMDL and IGMI. The second public hearings on the problems of chemical weapon destruction. Izhevsk. 1996. P. 83-86.
2. *Kolesnikova M.B., Yezhova N.N, Danilov G.E, KiryanovN.A.* The state of health of children living in the chemical agents stockpiling area. The second public hearings on the problems of chemical weapon destruction. Izhevsk. 1996. P. 93-95.
3. *Reshetnikov S.M., Ionov L.B., Makarova L.L., Zubcovsky N.E., Zakharov V.Yu.* The results of works on the complex investigation of the state of the environment. The second public hearings on the problems of chemical weapon destruction. Izhevsk . 1996. P. 51-58.
4. *Egidi D., Foraboschi F.P., Spadoni G., Amendola A.* The ARIPAR Project: Analysis of the Major Accident Risks Connected with Industrial and Transportion Activities in the Ravenna Area. Reliability

Engineering and System Safety, 1995, V.49, P. 75-89.

5. *Gorskij V.G.,Kurochkin V.K., Dyumaev K.M., Novocelcev B.N., Braun D.L.* Risk analysis is a methodical basis for the safety of chemical-technological objects // Rossijskij khimicheskij zhurnal. 1994. V. VIII. No. 2. P.54-61.
6. *Gorskij V.G., Polekhina O.V., Lepikhina E.A.* Models of toxic effects. GosNIIOKHT, 1994, 87 P.
7. Guidelines for Chemical Process Safety, American Institute of Chemical Engineers. N.Y., 1989, 585 P.
8. *S. Franke, F. Koehler, H. Zadach.* Chemie der Kampfstoffe, Teil 1 (CH-3700 Spiez, 2 Juli 1996/LYA-WO)
9. Chemical and bacteriological(biological) weapons and the effect of their possible use. Report of secretary-general. New York, United Nations, 1969.
10. Chemical and biological weapon today, SIPRIT2, Problems of chemical and biological weapon, V. 6, 1973
11. *Karlson E., Konberg M., Runn P., Viner S.* Evaluation of the consequences of possible accidents at the object of storage of Lewisite in the Kambarka area // Rossijskij khimicheskij zhurnal. 1995. V. IX. No. 4. P.79-88.
12. *Covello V.T., Merkhofer M.W.* Risk Assessment Methods Approaches for Assessing Health and Environmental Risks,N.Y.,London, Plenum Press, 1993, 318 P.
13. Risk analysis of the chemical weapon stockpile in Kambarka, Udmurt Republic, Russia. / Edited by Kolodkin V.M. Izhevsk. Publ.House of USU. 1995. 113 P.
14. Risk analysis of the chemical weapons stockpiles located on the territory of the Udmurt Republic, Russia. / Edited by Kolodkin V.M. Izhevsk. Publ.House of USU. 1996. 219 P.
15. *Bekker U.* Explosion phenomena:evaluation and consequences. ., 1996. 441 P.
16. *Korobejnikov Yu.V.* Gazodinamicheskie funkcii tochechnogo vzryva. ., 1969. 20 P.
17. *Dvojnishnikov A.C., Dorofeev C.B.* Eksperimentalnoe modelirovanie diagrammy razrushenija rezervuarov pod dejstviem vozdushnoj UV. Fizika gorenija i vzryva. 1992. No. 3. P. 112-116.
18. *Landau L.D., Livshic·E.M..* Teoreticheskaja fizika. V. 7. Teorija uprugosti. .: Nauka, 1987. 246 P.
19. *Astapenko V.A., Koshmarov Yu.A. Molchadskij I.S., Shevjakov A.N.* Termogazodinamika pozharov v pomeshchenijakh. .: Strojizdat, 1988. 415 P.
20. *Kolodkin V.M., Ionov L.B., Lebedev V.G. et al.* Shchuchye Arsenal of chemical weapons. Preliminary analysis of the threat to the population. /Kurgan. 1997. 40 P.
21. *Kurochkin V.K., Petrunin V.A., Sitnikov V.B., Fokin E.A., Shulga V.Ja., Safronov V.Ja.* Terrorizm s primeneniem khimicheskogo oruzhija // Toksikologicheskij vestnik. 1997. No. 3, P.11-17.
22. Problemy unichtozhenija khimicheskogo oruzhija. Sbornik nauchnykh trudov N 1. Ministerstvo oborony SSSR, Vojskovaja chast 52688- . , 1990, 108 P.
23. *Kholstov V.I., Tarasevich Yu.V., grigoriev S.G.* Puti reshenija problemy bezopasnosti ob'ektov po unichtozheniyu khimicheskogo oruzhija // Rossijskij khimicheskij zhurnal. 1995. V. IX. No. 4. P. 65-73.
24. *Kolodkin V.M., Morozov O.A., Danilov S.Ye.* Principles of building of the computer system forecasting accident consequences in storage, transportation and destruction of chemical weapon.// Izhevsk. Vestnik USU. 1994. P. 83-86.

PUBLIC INVOLVEMENT: MATTERS FOR THE PEOPLE IN TECHNOLOGY APPLICATIONS

JUDITH M. ESPINOSA, Director
ATR Institute
University of New Mexico

After the general presentation, the questions began at the back of the room. What appeared to be an older gentleman, stood to ask the technology expert to explain how the equipment could burn the plastics and other materials without causing bad air pollution problems. The expert answered in a somewhat perfunctory manner how this was an innovative technology which has been used in other countries. It was a closed loop system that did not allow for the escape of any harmful gases including dioxins. He then went on to repeat some of his previously prepared overview. Courteously but not daunted by what appeared to be an elusive technical explanation, the older gentleman continued with his questions. Was this an incinerator in the traditional sense? If not how does the material get "burned" or otherwise disintegrate? What happens to the toxins in the plastic? And, the by-products; is it ash, gases, liquids? Where will they go? As each question was answered, the rest of the audience seemed to pay more attention. They were simple questions, but it was becoming annoying for the scientific experts to get into so much detail. After all, people don't really need to know all the data on these new methodologies. The government agency had already reviewed the technology specifications prior to the public hearing, thus assuring everyone of its ability to perform.

Then others in the sparse audience began to join in for questions. Why hadn't they known before this about where this incinerator was to be located? Why does it have to go in their neighborhood? When was the decision made to put it here? How do we know if it is safe? At last the older gentleman requested that he and his neighbors be allowed to see the specifications and research data on this new technology application, so that they be better prepared to make a decision. Another public hearing was requested. The meeting was labeled as disruptive, and was adjourned.

The arguments commenced about whether the technology specifications should be publicly viewed. The media picked up on the story, and politicians and environmentalists came to the rescue of the community. After another attempt to permit the technology, it was deemed a failure. The company gave up. The government scientists and private sector technology experts were disappointed. They had reviewed and verified the data, and had visited another site where the technology had been utilized for a year prior. They were convinced of its appropriateness for permitting.

What was later learned was that the older gentleman was a retired engineer who had experience with some incinerator techniques over the years. He was not

F.W. Holm (ed.), Effluents from Alternative Demilitarization Technologies, 141–148.

antagonistic to the process at all, but felt it was his duty to have a good explanation for himself and his neighbors. He explained that after his questions were evaded, he himself became suspicious of the technology, and began to wonder about its efficacy. It was strange to him why he was not allowed to review the specifications, since he felt he could be useful to his neighbors.

This scenario is repeated time and again in many localities throughout the United States and from all indications in other parts of the globe. That is, despite the best scientific methodologies and research a technology can fail or be abandoned. In the majority of cases where the technology is otherwise feasible, the primary purpose for its failure can be found when the government agency or the technology experts lack the support of the public. Gaining public support is now archived in the literature as a collaborative process of "public involvement."

Public involvement is not to be confused with public information, public relations or other purely advertising or information dissemination tools. For the term *involvement* connotes an engagement, or connection, or active participation on the part of the person to whom the information is being imparted–in this case the public. At its best, *involvement* also means that the person on the receiving end of the information will accept the responsibility of becoming educated in the subject matter and prepare to participate in a meaningful communication and understanding. In a civil society, the institution that deals with the public on an issue of importance to the quality of its constituent's lives must be mindful of its responsibility to reach decisions based on effective participation.

PUBLIC INVOLVEMENT IS THE ABILITY OF AN INSTITUTION
TO ACHIEVE DIALOGUE WITH THE PUBLIC,
IMPART VALUABLE INFORMATION,
AND
CREATE NEW KNOWLEDGE WITHIN THE COMMUNITY

Figure 1.

Informed involvement at all levels of society is needed to increase the social capacity within a community, better the human development, expand economic opportunities, and improve the technology base on which society will come to depend in the coming decades. This is not an easy task, nor should it be expected to be, in an era where we see greater demands from the public, and various interest groups, to play a more profound participatory role in public policy development and governance. Regardless of the motives driving these demands, all institutions, including those from the scientific community, must respond efficiently and positively.

1. Expanding Access to Scientific Information

For decades environmentalists have demanded a stake in the outcome of the regulatory process and the methodologies used to prevent environmental degradation. Environmental laws have long made the assumption that citizen participation was a necessity in environmental sustainability and in curbing further pollution. Institutions that have dealt with environmental and natural resource issues know that public accountability is inherent in the system. While conflicts still exist in the implementation of environmental laws and regulations, the public is engaged, and it is only through its choosing that it can be excluded. This same scenario has not often been true for other public policy or decision making generally, and, most certainly, scientific research and technology applications have been somewhat isolated from public scrutiny.

Because scientific information is critical to the foundation of the knowledge needed to make accurate and informed life decisions, science must embrace public involvement at all steps in the scientific process. Public involvement is a policy of inclusion and as such serves to bring long term prosperity in a civil society. Scientific information and research is a key to that prosperity for increasingly we rely on basic scientific research to take advantage of opportunities from technological achievements.

The scientific community must begin to create an environment conducive to public involvement. By doing so this will strengthen the base of scientific knowledge and technology developments within our citizenry.

Without question any scientific research should produce objective results which can withstand scrutiny, not only from professional colleagues but also from the general citizenry. At times the most simple of inquiries from members of the public can serve to advise the scientific process, if scientists and technologists are open to the process. So public dialogue must not be seen as obstructive or simplistic even when the science seems complicated or obscure.

Figure 2: "Building a Framework for Scientific Understanding with the Public," depicts the cornerstones of how to build capacity within society for scientific discourse.

The growth of civil society's insistence on participatory governance throughout this decade will not be confined to any one particular entity. It pervades all quality of life issues which society faces, including the demand for good scientific information and responsible technology applications. The scientific community must be committed to the framework for public involvement by allowing expanded access to scientific information–however detailed and technical. That commitment should take the form of creating new methods of providing quality information to the public and other societal institutions. It behooves scientists to explore new means to manage data and improve the availability to and usefulness of their data to decision-makers. It is often these decision-makers, who will make policies which ultimately affect the work done by the scientific community. They are a critical link between the development of the technology and its application in the public arena.

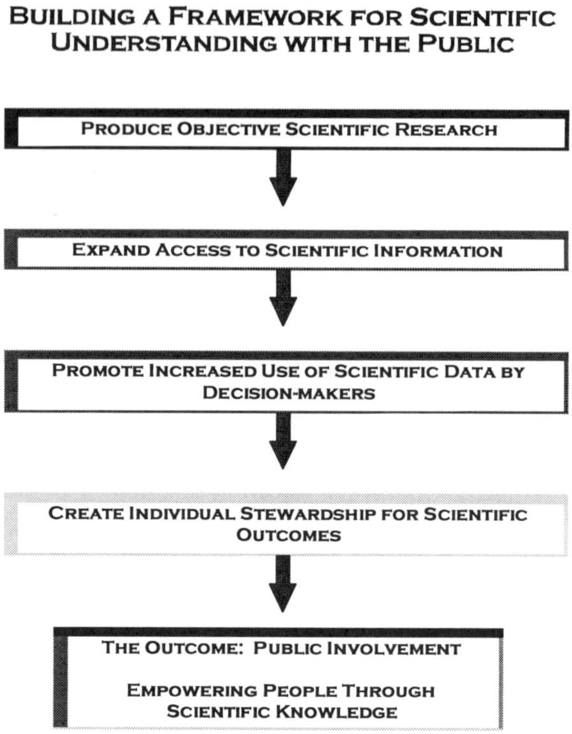

Figure 2.

2. Public Involvement in Technology Applications

Public involvement is a time consuming process that should not be viewed as a quick solution to getting a technology application out. It is a lengthier process than just imparting general information, because it is designed to allow for participation at every step in the scientific process. Figure 3: "Targets to Strengthen the Base of Scientific Knowledge," illustrates some of the steps that should be considered when seeking the public's participation in technology applications.

At a minimum the public should have a clear understanding of each step, and the procedures used to achieve a final product and process evaluation.
To the scientist in the public participation process the possibility of setbacks may seem formidable. It would be well to consider possible setbacks both before and during

Targets to Strengthen the Base of Scientific Knowledge

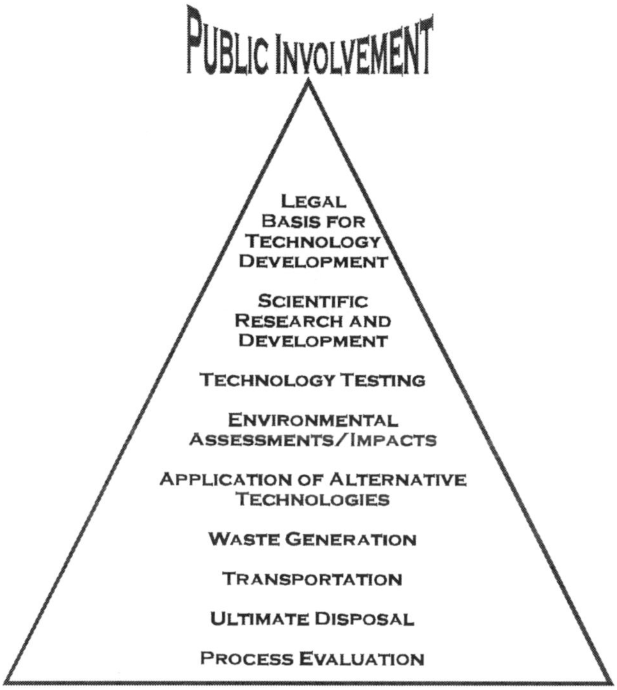

Figure 3.

active dialogue phases, and to be fully prepared to fashion them into part of the overall strategy of the technology application. Examples are,

- Increased information may lead to more disagreements
- Implementation phases of the process may be slowed
- The ultimate direction may change from that first anticipated
- Consensus may be more difficult to achieve
- Ideological differences may attempt to prevail

Flexibility during the public debate on a technology issue and the ability to control the course of the participatory proceedings are a must. Control does not lie in deciding who participates or what they choose to say. It is rather a control of the circumstances within which all citizens feel that they can actively take a role and be part of the decision-making.

A set of values by which to guide public involvement should be adopted simultaneously with the development of the technology. For eventually, that technology will be released for testing and use. Its acceptance by society will be directly related to the involvement the public had in bringing the technology to a credible application.

VALUE SETS FOR PUBLIC INVOLVEMENT
IN TECHNOLOGY APPLICATIONS

- Seek common goals
- Move from conflict to collaborative processes
- Determine public needs within the scientific process
- Emphasize sustainable technology applications
- Utilize modern communication techniques
- Promote integrity in the public dialogue
- Respect the diversity of people within the community

Figure 4.

3. Bringing Technology Applications to Communities

Technology is now a vital component of any community. It effects the quality of each citizen's life, and it can afford unprecedented educational, employment, economic, recreational, and environmental opportunities never yet experienced. It has improved productivity in the workplace and spurred the nation's economic growth. But lack of education in, and information about, technology and scientific research may make the public skeptical about their usage within their communities. The challenge is to meet the public involvement policy of inclusion that serves to bring long-term prosperity to a civil society. The challenge for a society that strives to be inclusive is to (1) find effective means for conveying detailed science and technology information to the populace, and (2) find new, more convenient, approaches to engaging the members of the public in planning decisions and defining what their needs are.

The ability of scientists to use innovative technology as a community development tool should be viewed as an opportunity in any public participation process. For example, advanced community-focused information systems can help address issues of access to research data and information that the public can use to analyze new technological applications. Using technology tools to educate the public on advanced technology applications should be fully explored along with traditional means of engaging the public. Community benefits to using technology include:

- Access to information
- Participating in the scientific process
- Development of local technology solutions
- Improved technology base
- Non-traditional means of public involvement
- Refined analysis for better solutions
- Creation of direct user interface

In the end, innovative technologies and research should be made applicable to a civil society, thereby preserving the nation's integrity and competition in the global community.

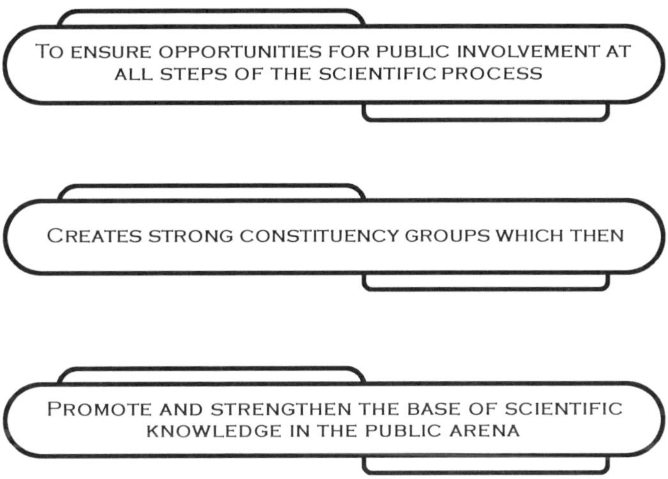

Figure 5.

If a civil society can build a strong public constituency reflective of the diversity of its people who will advocate for science and technology advancements, there will be great advantage to the scientific community. A greater vocal constituency actively involved in technology development and applications can only mean that the base of scientific knowledge is expanded into the general population, thereby creating more focus on these issues with policy decision-makers.

4. Bibliography

The following writers and reports provided the author with insights in preparing the presentation for the workshop and the subsequent paper. The author would also like to acknowledge Judith Holm, U. S. Department of Energy, National Transportation Program, for sharing her acumen and common sense over the years in the public involvement obligations of institutions to their constituencies.

1. Heaton, G.R., Banks, R.D. and Ditz D.W. (1994*) Missing Links: Technology and Environmental Improvement in the Industrialized World*, World Resources Institute, Washington, DC.
2. International Institute for Sustainable Development (1994) *Trade and Sustainable Development Principles*, IISD, Winnipeg.
3. Jasonoff, S. (1996) The Dilemma of Environmental Democracy, *Issues in Science and Technology*, pp. 63-70.
4. London, S. (1995) Building Collaborative Communities, *Pew Partnership for Civic Change*, Pew Charitable Trusts, Philadelphia.

5. National Research Council Committee on Risk Perception and Communication (1989) *Improving Risk Communication*, National Academy Press, Washington, DC.
6. National Research Council Policy Division (1996) *Linking Science and Technology to Society's Environmental Goals*, National Academy Press, Washington, DC.
7. Owen, S.D. (1994) Participation and sustainability: the imperatives of resource and environmental management, in R. Dobell and M. Neufeld (eds.), *The North American Institute: Trans-Border Citizens, Networks and New Institutions in North America*, Oolichan Books, British Columbia, pp. 45-52.
8. Putnam, R. (1995) Bowling Alone: America's Declining Social Capital, *Journal of Democracy*, pp. 65-78.
9. Ronfeldt, D. and Thorup, C. (1994) NGO's, civil society networks, and the future of North America, in R. Dobell and M. Neufeld (eds.), *The North American Institute: Trans-Border Citizens, Networks and New Institutions in North America*, Oolichan Books, British Columbia, pp. 21-39.
10. Sitarz, D. (eds.) (1993) *Agenda 21: The Earth Summit Strategy to Save Our Planet*, Earth Press, Boulder.
11. The President's Council on Sustainable Development (1996) *Sustainable America: A New Consensus for the Future*, U. S. Government Printing Office, Washington, DC.
12. The President's Council on Sustainable Development (1996) *Sustainable Communities Task Force Report*, U. S. Government Printing Office, Washington, DC.
13. Warth, A.H. (1992) Public Participation in Technological Decisions; A New Model, *Bulletin of Science, Technology, and Sociology*, pp. 289-339.

GREEN CROSS RUSSIA'S EXPERIENCE IN FORMING THE PROGRAM FOR JOINT ACTIVITY OF THE PUBLIC AND GOVERNMENT IN THE DESTRUCTION OF CHEMICAL WEAPONS

VLADIMIR LEONOV
Green Cross Russia, Program Director

As a non-governmental environmental protection organization, Green Cross Russia considers its main task to be attracting attention and funds from the world community and the Russian government to solve complicated national and international environmental protection issues.

One of Green Cross Russia's primary programs is the international program "Legacy of the Cold War." This program is aimed at exposing problems connected with degradation of the environment because of the arms race and the activities of the military-industrial complex, and at searching for ways to solve these problems. The program is divided into:

- Informing the world community about existing problems,
- Collecting, systematizing and analyzing information about pollution and worsening environ-mental conditions,
- Creating public awareness of the necessity for solving these problems,
- Conducting joint events of various institutes on development of solutions that help to overcome the harmful ecological consequences of the arms race,
- Taking practical measures to investigate the pollution of territories and the health of the population, and to rehabilitate them.

The primary directions of the activities of the program:

- Collecting, systematizing, and analyzing information on the technogenic pollution of territories,
- Conducting comprehensive work to explain to the population the problems associated with the destruction of chemical weapons,
- Studying polluted territories,
- Investigating the health status of the population,
- Conducting rehabilitation measures.

The preparation for the destruction of reserves of chemical weapons has a special place in the work toward overcoming of the consequences of the arms race.

Russia's preparation to ratify the Convention on the Prohibition of Chemical Weapons has entered the final stage. However, the work to prepare for their destruction is moving extremely slowly. A search for effective forms and methods of influence to

F.W. Holm (ed.), Effluents from Alternative Demilitarization Technologies, 149–153.

achieve the above task is necessary. Furthermore, safe technologies and working conditions must be created for the destruction of toxins.

The unfortunate experience of the construction of the plant for destroying chemical weapons in the city of Chapayevsk in the Samara Oblast in Russia showed that confrontation between the public and authorities is not a viable option toward a solution to this problem; since, despite the fact that both sides support fulfillment of the Convention on Chemical Disarmament, each side has its own view as to how the problem should be solved.

The reason for the development of the situation is many years of governmental arms race policies, which were conducted with the complete neglect of questions of ecology, a disregard for the health of hundreds of thousands of people, as evidenced by the absence of predictions of possible consequences to these people, and without public information.

The situation today, in a number of regions in which chemical weapons are to be destroyed, is dangerous, mainly, because it could lead to the repetition of the unfortunate experience of spending great effort and many material resources on confrontation rather than constructive steps -- even small ones -- directed toward solution of the problem of chemical disarmament.

At this time, as the first steps are being taken, according to the existing federal program for the destruction of chemical weapons, popular concern is growing and in certain areas is even turning into opposition. The main issues alarming people are an all but complete lack of information, the safety of the destruction processes, and the effects of chemical destruction plants on the environment.

In connection with this, Green Cross Russia offered to conduct public hearings on these problems. The public hearings could include an open discussion between representatives of the public at large and public organizations, the authorities, the Ministry of Defense, the Committee on Conventional Problems of Chemical and Biological Weapons, medical personnel, ecologists, and the developers of technologies for the destruction of chemical weapons. This is an entirely new element in the cooperation between the public and Russian power structures in this sphere.

Green Cross Russia proposed to conduct public hearings under the title "From Confrontation to Cooperation," the primary goal of which would be to try to establish mutual understanding between the population in general and public environmental movements on the one hand and power structures and departments on the other, and to move forward to practical measures for the solution of this complicated ecological problem.

The given initiative received the approval of president Boris Yeltsin, the support of local governmental authorities, concerned ministers, the state Duma of the Russian Federation, and society. The first hearings were held in Saratov from October 17-19, 1995, the second in Lzhevsk and Kambarka from May 13-17, 1996, and the third in Kurgan and Shchuch'ye from July 7-10, 1997.

The goals of the hearings were:

- open and free discussion of the entire complex of political, ecological, social, medical, economic, technological and other problems connected with the destruction of chemical weapons,

- the highest possible level of awareness of the population about the state of the given problems, and ecological education of the population at large and representatives of regional administrations,
- reduction of social tension, and liquidation of confrontation between the population and the authorities,
- unification of the efforts of all sides to achieve the safe destruction of chemical weapons,
- strengthening of international cooperation on the problems of the prohibition and safe destruction of chemical weapons.

At the hearings, presentations were given by government representatives and specialists in various aspects of the problems involved in destroying chemical weapons and rehabilitating the health of the population and the environment.

At the most recent hearings in Kurgan, 67 participants presented reports or information. Included among these participants were representatives of presidential and legislative structures, delegates of the state Duma of the Russian Federation, representatives of the administration and population of the Kurgan Oblast and the Shchuch'ye district; representatives of ministries, departments and committees, scientists and specialists developing documentation for the construction of a plant for destroying chemical weapons, and documentation for development of the infrastructure of the district in which the construction is to take place; and representatives of the government contractor companies and society of the USA. In addition, 20 papers, reports and other communications were displayed.

The hearings were preceded by extensive work to explain their necessity to the Ministry of Defense and the primary departments involved in the process of destroying chemical weapons. A series of preparatory meetings with the leadership and population of the Kurgan Oblast were held. An important part of the preparation was work done by independent groups of specialists to understand the situation in the region and to educate the population. This work included:

- the establishment of informational, outreach offices in Kurgan and Shchuch'ye on the problem of the destruction of chemical weapons,
- the inspection of the health status of the individual groups of the population, and the status of medical help available to the population in the Shchuch'ye district of the Kurgan Oblast,
- the development of methods of health prognosis and management based on study of the cause and effect relations between hygienic factors, living conditions and people's health status,
- the development of methodological approaches to monitoring the health of people exposed to anthropogenic factors, including extremely minute concentrations of toxic substances,
- a comparative risk assessment of the stockpiling and destruction of chemical weapons,
- the special training of medical personnel of district hospitals in areas of storage of combat toxins, focusing on medical-hygienic and ecological safety aspects during the destruction of chemical weapons,

- ecological education seminars for representatives of the government,
- seminars for ecological education of the population,
- the publication of literature to educate the population, e.g., popular brochures.
- mass media campaigns.

In addition, 200 children from all seven regions of storage and proposed destruction of chemical weapons sites were brought to a sanatorium in an ecologically clean region of Russia for the examination and development of recommendations for rehabilitation of health and individual adaptation in actual living conditions. Among these children were 30 individuals from the Shchuch'ye district of the Kurgan Oblast.

In the course of the hearing a final document, which expressed a large number of desires of the population of the Kurgan Oblast directed to legislative and governmental bodies, was produced and passed.

Results of this work were quickly evident. These include:

- the realization that the work with the population to explain the problem of chemical disarmament was begun very late,
- a gradual lessening of social tension and a transition to coordinated action of the population and the authorities is taking place, moving from confrontation to cooperation, for example, after the hearings many questions were entertained by the information center as to what kinds of specialists will be needed for the construction and work of the plant for destroying chemical weapons, and how jobs can be obtained,
- a group of independent experts that is currently examining the first documentation for the plant for destroying chemical weapons;
- large advances that have been made by the Ministry of Defense to organize work with the population on the given problem.

Today it is possible to recommend the practice of organizing and conducting public hearings as one of the most important instruments for resolving situations of conflict in Russian society, such as the situation surrounding the problem of destroying chemical weapons. The hearings must be well prepared for and supported by preliminary work with the population and legislative and state institutions. The hearings will sharply raise the level of awareness of the population and the overall level of ecological education of the population, and will lower the level of social tension.

It is necessary to invite a wide array of public organizations, including radically-minded ones, in order to understand their positions and involve them in constructive mutual action.

It is recommended that an assessment be obtained for the population of the regions used for the storage of chemical weapons regarding the risk of accidents occurring at the storage sites versus the risks during the destruction process. It is also necessary to conduct independent medical evaluations of the population, especially woman and children. The goal of such evaluations is to detect background patterns in the development of diseases found in members of the population and the presence (or absence) of influence by chemical weapons on disease dynamics.

The recommendation to conduct independent ecological impact studies of the project for the chemical weapon destruction facility could be important for upholding the right of residents to guaranteed safety during the process of destroying chemical weapons.

The recommendations of the hearings that were recorded in the concluding document must influence delegates of the state Duma and help them resolve the question of ratification of the Convention on the Prohibition and Destruction of Chemical Weapons.

It is clear that the cooperation of the authorities, representatives of industry and the population must be constant -- for all periods, including storage, preparation for construction of a chemical weapons destruction facilities, and the destruction process itself. Timely education of the populace about the possibilities of converting a plant after chemical weapons have been destroyed is also important.

One of the most recent events organized by Green Cross Russia was the action concerning the necessity to ratify the Convention on the Prohibition and Destruction of Chemical Weapons. A letter to the state Duma was produced and sent to a great many public organizations with requests for support. It is clear that chemical weapons must be destroyed. It is also clear that the population must be provided with guarantees of safety during the destruction process that apply to both people and the environment. Nevertheless, the Convention still has not been ratified, and nothing further is being accomplished. Many organizations supported the letter. Green Cross Russia plans to present public opinion on the question to the state Duma for examination as one of the important arguments in favor of ratification of the Convention.

RESEARCH NEEDS

VINUTHA S. GOWDA
SAIC, 2109 Emmorton Park Road, Edgewood, Maryland

FRANCIS W. HOLM
30 Agua Sarca Road, Placitas, New Mexico

A panel was assembled at the end of the Advanced Research Workshop to deliberate the additional research needs in the area of Effluents of Alternative Demilitarization Technologies worthy of special mention. The discussants examined the current technologies and policies that were presented by each of the speakers and were able to reach a consensus over future research needs. The workshop participants developed the following seven recommendations:

1. There is a pressing need to rigorously characterize the products of incomplete reaction (PIRs) in effluents of alternative demilitarization technologies.

2. In the use of biodegradation technologies, it is especially important to characterize the by-products of reaction and their partitioning in bioprocessing effluent streams.

3. Since heat and mass balances are generally not available for alternative technologies, there is an urgent need to develop, at least, mass balance data on all alternative demilitarization technologies including PIRs (trace elements) and by-products.

4. The formation of reactive or toxic by-products (CO, H_2S, H_2, etc.) in alternative technology effluent streams, brings an urgent need to perform detailed process safety reviews and ultimately total quantitative risk assessments.

5. Although real-time monitoring in gaseous effluent streams is relatively mature, there is a pressing need to develop continuous real-time monitoring for liquid effluent streams for process control and for controlling toxic substances.

6. Depending on the technology, there remains a need to provide management and control of dioxins/furans (and their precursors) in all waste streams.

7. For the introduction of new alternative technology to a site-specific application, methods need to be established for engaging the public and technical community in the process of technology selection, risk assessment, waste minimization, and reducing the potential environmental burden.

F.W. Holm (ed.), Effluents from Alternative Demilitarization Technologies, 155–158.

1. Characterizing the Products of Incomplete Reaction (PIRs) in Effluents of Alternative Demilitarization Technologies

Alternative demilitarization technologies are grouped into three categories: low, moderate, and high temperature processes. Future analysis for these technologies should focus on an accurate characterization of gaseous, liquid, and solid effluents for products of incomplete reactions (PIRs). A comparative analysis between PICs of incineration and PIRs of alternative demilitarization technology would be an additional measure of the effectiveness of an alternative technology.

Operating permits needed for treatment units will likely result in a more lengthy permitting process. It is likely that some new sampling and analytical methods will need to be developed for the Products of Incomplete Reaction (PIRs) generated in the new technologies which may be present in the air, water, and solid effluent from these processes. The PIRs from acid digestion processes for example, could contain nitrated volatile and semi-volatile.

2. Characterizing and Partitioning the By-Products of Biodegradation Technologies

In the use of biodegradation technologies, it is especially important to characterize the by-products of reaction and their partitioning in bioprocessing effluent streams.

In the area of bioprocessing, characterization of the effluent waste stream by organic composition, metal content, and oxidation products should be fully developed. A mass balance is needed on the bioprocessing unit detailing the amount of carbon entering the reactor to the CO_2, cell mass, and organic by-products that are leaving in the effluent stream. The conditions required for each reactor, anaerobic or aerobic, should be fully explained and characterized.

3. Development of Mass Balances for Alternative Demilitarization Technologies

Since heat and mass balances are generally not available for alternative technologies, there is an urgent need to develop, at least, mass balance data on all alternative demilitarization technologies including PIRs (trace elements) and by-products.

For alternative technology effluents, a complete mass balance on all components should be provided. All components in the initial feed stream should be tracked to the point of destruction or waste in the effluent stream. A destruction efficiency analysis should be preformed for a comparative analysis with other alternative technologies and baseline technologies. Identification and quantification of all raw materials, process intermediates and product streams for gas, liquid, and solid emissions is needed to fully understand the process.

4. The Formation of Reactive or Toxic By-Products

The formation of reactive or toxic by-products (CO, H_2S, H_2, etc.) in alternative technology effluent streams, brings an urgent need to perform detailed process safety reviews and ultimately total quantitative risk assessments.

A safety analysis is needed to address what hazards will be introduced from the process through raw materials, operating procedures, products, and waste streams. A qualitative hazard risk assessment for 30, 60, and 90% design reviews should be established to include issues such as worker interactions during operations, personal protective clothing requirements, safeguards, containment precautions, and monitoring needs for specified detection levels. This established methodology should be utilized for the assessment of technology, the examination of technology effluents, and the development of the technology into a viable option in comparison to incineration. It should also be tested to detail potential accident scenarios, contingency planning, and a preliminary hazard analysis.

5. Continuous Real-Time Monitoring

Although real-time monitoring in gaseous effluent streams is relatively mature, there is a pressing need to develop continuous real-time monitoring for liquid effluent streams for process control and for controlling toxic substances.

Near real time characterization of PICs and Volatile Organic Carbon (VOC) that are released from processing chemical agent or other toxic components should be included in any assessment of alternative technologies. Trial runs should be used in a preliminary risk assessment of mass balances to track and monitor near real time effluents of toxic metals as particulate matter and VOCs. Analysis of solid, liquid, and gas emissions are needed to verify compliance with environmental regulations.

The local government authorities are likely to consider the use of the waste characterization requirements for waste incinerators for regulating high temperature alternative technologies because of the large experience base. This might not be appropriate for medium and low temperature alternative technologies; consequently, additional work will probably be required in developing more appropriate waste characterization parameters.

It is likely that low temperature processes will have higher concentrations of pollutants and PIRs in the water phase than the high temperature treatment processes. This is an important process development issue because the traditional procedures are not as well suited to monitor and control trace quantities of PIRs in the water discharge. By comparison, trace Products of Incomplete Combustion (PICs) in the stack gas of incinerators are well understood. A significant level of effort may be required by local governments to develop monitoring, analytical, and control procedures for the aqueous effluents from the low temperature treatment processes.

6. Control of Dioxin and Furans and Their Precursors

Depending on the technology, there remains a need to provide management and control of dioxins/furans (and their precursors) in all waste streams.

Management of stack emissions is needed to reduce releases of chlorinated dioxins, furans, chlorinated phenols, and benzene in effluent streams of incineration for the health and safety of working environments. New research technologies should monitor and control emissions of these toxic chemicals into public waste streams. Alternative reduction methods should be investigated in order to manage this risk to acceptable levels.

7. Engaging the Public and Technical Community in Process Implementation

For the introduction of new alternative technology to a site-specific application, methods need to be established for engaging the public and technical community in the process of technology selection, risk assessment, waste minimization, and reducing the potential environmental burden.

Host communities have expressed concern about potential long-term health hazards and environmental burdens from emissions of an agent or toxic compound and its reaction by-products. A better mechanism for public exchange is needed to mitigate these concerns. A public forum should be established to discuss methods, status of current design, and implementation of the technology. The forum should act as a conduit to encourage participation between the government, industry, and the public.

The characterization of the effluent streams of alternative demilitarization technologies was a prevalent theme throughout the Advanced Research Workshop. This characterization is crucial to the success of the technology as well as the safety and health of the surrounding community. These research needs should propel future development of the alternative technologies in a manner that is aimed at improving the acceptance of the technologies to the general public and environmental regulators.

APPENDIX: DESCRIPTIONS OF ALTERNATIVE DEMILITARIZATION TECHNOLOGIES AND ESTIMATED MASS BALANCES

FRANCIS W. HOLM
30 Agua Sarca Road, Placitas, New Mexico

1. Comparisons

Preliminary mass balances are useful to begin to assess how to dispose of process effluents. Until relatively large-scale tests have been performed, it is not possible to accurately predict the concentrations of trace contaminants in the effluents. It will be these contaminants, and their level of concentration, that will determine if the effluent is ready for ultimate disposal or if further treatment is required. Further treatment could either involve more process equipment or shipping the effluent offsite to a commercial waste treatment facility. More detailed information is presented in a recent report entitled "Alternative Technologies for Destruction of the Chemical Weapons Stockpile," F.W. Holm, Final Report for the U.S. Army, Available from the Document Control Center, Aberdeen Proving Grounds, MD, November, 1996.

Material balances in this report are based on vendor recommended operating conditions and stoichiometry. The resulting mass balances are calculated to conserve elemental mass flow into and out of each technology. The results are not based on actual destruction data for chemical warfare agent, except for the five technologies evaluated by the NRC, where detailed conceptual design packages were available. For these five technologies, neutralization, biodegradation, electrochemical oxidation, catalytic extraction process (also referred to as molten metal process) and gas-phase chemical reduction, two significant figures after the decimal point are used in the tables throughout this section to show increased confidence in the estimated values.

A summary of effluents from all the individual technologies presented in this section is presented in Table 1. The estimates of quantities of effluents in Table 1 are theoretical and could be impacted greatly by actual operating procedures. Some of the operating parameters that could increase the quantity of effluent produced include: excess reactant chemicals, excess air or oxygen, excess diluents for pH control, excess scrubbing fluids, and excess quenching fluids.

2. Molten Metal (Steel, Nickel, or Copper) Alloy

The molten metal technology uses a high-temperature pool of liquid metal to dissolve and extract recoverable constituents, both organic and inorganic (see *Figure* 2 and Tables 2a through 2c). Waste materials are injected into the molten metal bath, which is the reaction medium for the production of specific byproducts. Large molecules, specifically those found in chemical warfare agents, are thermally decomposed (elemental dissolution) in the metal bath at 2,912° to 3,092°F (1,600° to 1,700°C) to

F.W. Holm (ed.), Effluents from Alternative Demilitarization Technologies, 159–213.

TABLE 1. Estimated Mass of Effluents for Technologies Destroying Chemical Agent[a]

Technology Agent Treated	HD			VX			GB		
Effluent Phase	Gas	Liquid	Solid	Gas	Liquid	Solid	Gas	Liquid	Solid
Molten Metal (oxygen)	1.19b	1.46	0.21	1.73b	1.23	0.23	1.0	2.0	1.4
Carbon Electrode (air)	1.6	1.7	1.3	3.7	2.6	0.7	2.5	1.7	1.2
Plasma Torch (air)	1.6	1.7	1.3	3.7	2.6	0.7	2.5	1.7	1.2
Induction Coupled Plasma (air)	1.6	1.7	1.3	3.7	2.6	0.7	2.5	1.7	1.2
Molten Salt Oxidation	16.7	1.6	1.4	24.7	4.9	0.2	17.5	2.2	3.0
Gas Phase Reduction with Hydrogen	0.77	2.22	...	0.74	0.46	0.55			
Hydrotreating	1.2	1.5	1.8	1.6	2.6	1.3	1.4	1.7	1.3
Supercritical Water Oxidation	1.4	23.1	...	2.3	25.5	...	1.6	22.1	...
Wet Air Oxidation	6.2	17.3	...	12.2	30.0	...	6.5	21.0	...
Steam Reforming	6.4	...	1.6	11.9	...	1.2	8.6	...	1.2
Reaction with Sulfur	0.0	3.3	0.5	0.7	0.7	1.3	0.7	2.0	0.8
Electrochemical Oxidation with Silver	4.11	13.84	6.15	6.72	23.7	8.63			
Water Hydrolysis	1.65	29.53	...						
Biodegradation of Hydrolysis Products	67.94	92.90	...						
Caustic Hydrolysis				1.73	11.04	...	...	3.9	...
Catalytic Oxidation	19.8	...	0.1	20.0	...	0.1	19.7	...	0.1
Thermal Media Oxidation	19.8	...	0.1	20.0	...	0.1	19.7	...	0.1
Gasification	1.6	1.6	1.4	3.5	2.8	0.7	2.5	1.9	1.2

a Pound of effluent per one pound of Agent destroyed

b Values with two significant figures after the decimal point are based on agent destruction data reported by the respective vendors.

small molecules or atoms. An example byproduct from this process includes a fuel gas stream containing nitrogen (N_2), hydrogen (H_2), carbon monoxide (CO), and a solid residue best characterized as a vitrified ceramic (a glass-like material) or slag.

Decontamination of drained ton containers would be accomplished in an indirectly heated premelting chamber located on the side of the reactor. Drained ton containers would be smelted in the side chamber allowing the premelted metal to drop into the molten pool inside the reactor. A sweep gas also carries the offgas into the reactor. The metal is periodically tapped from the reactor and cooled to form an ingot for recycle.

The system uses two catalytic processing units. The primary and secondary catalytic processing units operate at 2,700°F (1,482°C) at different pressures, 1 atmosphere [14.7 pounds per square inch actual (psia)] and 2 atmospheres (30 psia), respectively. All bulk solids, including ton containers, dunnage, and agent residue, are processed in the primary unit. The secondary unit receives the gas stream from the primary unit and serves as a polishing reactor to ensure offgas quality. The nerve agent (VX) drained from the ton containers would be injected into the primary unit through tuyeres (inlet piping) underneath the pool of molten metal. The mustard agent (HD) drained from ton containers would be injected into the higher pressure secondary unit, also through tuyeres in the bottom of the molten metal pool. Periodic metal tapping is expected to produce a metal product with approximately 9 weight percent (wt %) of both sulfur and phosphorus.

In the gas handling train, the offgas is cooled with methane and processed through a high-temperature ceramic filter. The filtered solids are recycled to the reactors and the gas is cooled further by heat exchangers. Elemental sulfur, also a solid, is removed from hydrogen sulfide (H_2S) in the gas train. The only liquid effluent expected is a 17 wt % aqueous hydrochloric acid (HCl) stream scrubbed from the offgas in the gas train. A synthesis gas exits the gas handling train composed of mostly CO and approximately 5 wt % H_2. The gas produced could be used as a heat source or for methanol synthesis.

A molten metal system to treat HD is being evaluated for operation at Aberdeen. On completion of stockpile destruction at Aberdeen, the equipment would be decontaminated and moved to Newport, where VX would be destroyed.

3. Molten Aluminum Alloy

The molten aluminum process uses a liquid alloy bath, composed primarily of aluminum, to breakdown organic chemicals and to capture metals introduced with the feed. The molten aluminum alloy reactor is indirectly heated using auxiliary fuel, such as natural gas; however, electric power could also be used. The reactor freeboard is purged with a sweep gas to exclude oxygen and create a reducing environment.

Waste organic chemical bonds are broken by the pool of molten aluminum alloy operating at temperatures of 1,472° to 1,652°F (800° to 900°C). Some of the gaseous products of decomposition are H_2, CO, ammonia (NH_3), and lower alkenes.

There are no commercial units operating on a day-to-day basis. Three prototype units have been built and tested: a soil treatment unit, a biomedical waste treatment unit, and an existing test facility. The soil and biomedical units have been abandoned. The test facility has multiple feed capabilities but is limited in capacity to small laboratory research size samples.

Some of the major knowledge gaps associated with this technology are:

- expected life and recycling potential of the alloy bath due to metal accumulation
- ultimate fate of chlorine in the reactor
- non-metallic slag removal from the molten alloy bath.

Data for this technology are insufficient to support an estimate of potential process effluents.

4. Molten Zinc, Tin, or Lead Alloy

Although no specific equipment or commercial application of this technology exists, it has been recommended as a molten metal pyrolysis process. The relatively high specific heat capacity of zinc, lead, and tin could be beneficial in reducing the size of the reactor, while maintaining efficient destruction. No detailed information was provided for this technology; therefore, mass balances could not be estimated. Some of the major gaps associated with this technology are:

- expected life and recycling potential of the alloy bath due to metal accumulation
- ultimate fate of chlorine in the reactor
- non-metallic slag removal from the molten alloy bath
- toxic metal emission control.

Data for this technology are insufficient to support an estimate of potential process effluents.

5. Carbon Electrode Technology

A carbon electrode reactor has two high-temperature reaction zones (see *Figure* 3 and Tables 3a through 3c). One is the general furnace atmosphere or freeboard zone of the furnace, which may be at a temperature of 3,200°F (1,760°C), depending on the chosen operating conditions. The second is the plasma zone (the arc between the carbon electrode and the molten metal) in which temperatures exceed 10,000°F (5,537°C) and may approach 25,000° to 30,000°F (13,871° to 16,648°C). Chemically, the furnace atmosphere can be controlled to be oxidizing, reducing, or neutral. It operates at a higher temperature than most combustion-based incinerators; however, it does not depend on exothermic combustion reactions to maintain its operating temperature. Also, the volume of gases in the furnace atmosphere is not dominated by burner combustion products and the associated high volumetric flow rates of reactants. The system offers close control of offgas composition and flow rate.

The reactor's main driving potential is the plasma zone, technically referred to as high-temperature thermal plasma. It is characterized by high viscosity, extremely high heat transfer rates, and molecular species that are predominantly ionized. The plasma is electrically neutral with an equal number of positive- and negative-charged ions present. It is electrically conductive and once formed, the plasma is stable. Chemically, large molecules are broken down into small fragments and ionized, and the plasma incorporates simple monatomic and diatomic ions (one or two atom species). Typical reaction products upon cooling to furnace temperature are: N_2, CO, HCl gas, hydrogen fluoride (HF) gas, H_2, phosphorus pentoxide (P_2O_5), oxygen (O_2), and carbon dioxide (CO_2). Depending on the conditions present in the furnace atmosphere, some oxides of nitrogen may also form.

Three electric arc systems have been tested and demonstrated. One facility, designed and tested by the Exide Corporation, is for treating contaminated soil. Exide's experience in lead-acid battery manufacturing and secondary lead smelting facilities was applied to treating contaminated soil at a former battery recycling operation. The site had a mixture of broken rubber battery cases in soil that was also contaminated with lead and lead compounds. Some success was shown on the results from field tests on smelting lead and destroying battery casings performed by the Exide Corporation in the US Environmental Protection Agency's (EPA's) November 1994, Superfund Innovative Technology Evaluation Program.

The second is a graphite electrode direct current arc system, demonstrated in both the Mark I, 0.3 MW furnace (Surma et al., 1993) and the larger Mark II, 1 MW furnace (Surma et al., 1994). These furnaces were jointly developed and demonstrated by Pacific Northwest Laboratory (PNL), Massachusetts Institute of Technology (MIT), and Electro-Pyrolysis, Inc., with funding support from Department of Energy (DOE). These systems are now being marketed by Kennedy Van Saun, Danville, Pennsylvania, a unit of Sweden's Svedala Industries, Inc. (Trescot et al., 1995).

The third electric arc system is the alternating current graphite electrode arc melter. Development of this system has been supported by DOE and is an extension of a U.S. Bureau of Mines (USBM) and the American Society of Mechanical Engineers (ASME) demonstration of the vitrification of municipal waste combustor residues. The furnace, developed in cooperation with Lectromelt, a subdivision of the Salem Furnace Company, is a sealed, 800 kVA (0.8 MW) arc furnace. It is generally operated between 350 and 550 kW power.

6. Plasma Torch Technology

A plasma torch design uses a refractory-lined vessel heated by a plasma torch. A carrier gas flows through the torch (air, oxygen, nitrogen, etc.) and becomes a high-temperature ionized plasma gas. Although centerline plasma temperatures may reach 21,632°F (12,000°C) or more, plasma surface and surrounding gas temperatures vary between 2,732°F (1,500°C) and 9,572°F (5,300°C). When air or oxygen is used as the plasma gas and a secondary steam is added as an oxidant and source of hydrogen, the reactor can produce a fuel or synthesis gas. The plasma reactor can process dunnage, laboratory wastes, and spent granular activated carbon (see *Figure* 4 and Tables 4a through 4c).

The plasma torch systems volatilize and decompose organic materials and melt inorganic materials into a glassy slag and, in some cases, into a separate molten metal phase. When withdrawn, the slag forms a leach-resistant vitrified (glassy) waste that is suitable for disposal. The molten metal phase, if formed, can be withdrawn separately and recycled for alternate uses of the recovered metal. To control viscosity and the characteristics of the molten ceramic slag that floats on top of the molten metal inside the reactor, sand (SiO_2), calcium oxide (CaO), or other conditioning chemicals can be added. A well-controlled ceramic product may be able to pass EPA toxicity characteristic leaching procedure (TCLP) criteria as an inert product.

In a plasma torch reactor, agent would be injected into the reactor once operating temperatures are reached. Although the plasma temperature is extremely high, intimate mixing between the liquid agent and plasma gas may not occur instantly, because the fluid dynamics of a fully formed plasma column resists mixing; however, the liquid will vaporize and pyrolyze due to thermal radiation from the torch and reradiating refractory. Some designs use a hollow cathode, which allows mixing before the plasma column is fully formed.

Because the very high temperatures that chemical agents and other waste materials are exposed to in the reactor, there is little concern that the products and process residues will not meet RCRA requirements and land disposal restriction (LDR) standards. Key effluents needed to meet regulatory requirements are the baghouse discharge that may contain toxic metals, particulate, and partially reacted lime-containing compounds of S, P, and Cl. Also, the use of product synthesis gas directly to generate process heat may present permitting problems; however, it does not appear that this technology will present unique permitting problems.

The US Department of Energy (DOE) has sponsored development, test, and evaluation of the Plasma Hearth Process (PHP), a nonradioactive, fixed hearth, plasma torch system at both the laboratory and pilot scale. A pilot-scale, nonradioactive PHP system will be constructed at Retech, Inc., in Ukiah, California. A bench-scale radioactive PHP system is being installed in the Argonne National Laboratory West (ANL-W) Transient Reactor Test (TREAT) facility in Idaho (U.S. Department of Energy, 1995).

Retech, Inc., has also developed the Plasma Centrifugal Furnace (PCF), a rotating-tub (hearth), plasma torch system. Several sizes, ranging from 45.7 cm (1.5 feet) to 244 cm (8 feet) in diameter have been tested. A 183 cm (6 foot), 136 kg/h (300 lbs./h) design has been used in DOE-sponsored tests at the Component Development and Integration Facility in Butte, Montana, to demonstrate its potential application in the treatment of mixed waste sludges at the Idaho National Engineering Laboratory (INEL). A system was installed in Switzerland in 1990, and Retech is delivering three more systems to European customers (Eschenback et al., 1993).

Other plasma torch systems are at various stages of development. INEL is developing a steam plasma torch process that treats liquid and gaseous wastes (U.S. Department of Energy, 1994c). Westinghouse Electric Corporation, under Electric Power Research Institute (EPRI) sponsorship, has developed a plasma cupola technology that is now being marketed by Westinghouse (Dighe et al., 1991).

7. Induction Coupled Plasma Torch

The induction coupled plasma torch is based on the concept of gas ionization by inducing an oscillating magnetic field in a flowing gas stream. The inductively coupled plasma (ICP) torch functions in a flowing gas medium, while the electric arc functions in a static gas medium. Although centerline plasma temperatures may reach 21,632°F (12,000°C) or more, plasma surface and surrounding gas temperatures vary between 2,732°F (1,500°C) and 9,572°F (5,300°C). Heat transfer to the waste material in the main reactor hearth is primarily by radiation with some contribution by convection from the surrounding gas. In addition, joule heating occurs when the waste material is used as one of the electrodes. In general, these systems can accept a variety of input wastes materials, even "as received" in their original containers.

The treatment process uses the induction coupled plasma torch as its heat source; the torch is coupled to a reaction chamber that treats gaseous, liquid, slurry, or solid waste streams. The plasma torch volatilizes and decomposes organic materials and melts inorganic materials into a glassy slag and, in some cases, into a separate molten metal phase. When withdrawn, the slag forms a leach-resistant vitrified (glassy) waste form that is suitable for disposal. The molten metal phase, if formed, can be withdrawn separately and recycled for alternate uses of the recovered metal. The anticipated products of this process depend on the waste form but could include methanol, acetic acid, HCl, ammonia-based fertilizers, gravel, and energy.

Alliant Techsystems and Plasma Technology, Inc., Santa Fe, New Mexico, is commercializing the Plasma Energy Recycle and Conversion (PERC) process that converts liquid and slurry wastes into useful products (Persoon, 1994).

8. Molten Salt Oxidation

Molten salt oxidation is a process in which organic chemicals, such as HD and VX, undergo thermal decomposition in a catalytic pool of molten salt (see *Figure* 5 and Tables 5a through 5c). The process uses a molten salt, sodium carbonate alone or mixed with potassium or lithium carbonate, as the salt for the bed media. Air and liquid agent would be introduced into an alumina brick-lined vessel containing the molten salt that is controlled by operating temperatures maintained between 1,292°F (700°C) and 1,742°F (950°C). Maximum solid and slurry particle size must be controlled to 3 mm (1 inch) for pneumatic conveying. In the molten bed the agent reacts with sodium carbonate and oxygen from the air to form CO_2, H_2O, and salts, which are formed from the acidic components of HD or VX. The reaction products of the HD are CO_2, H_2O, sodium chloride, and sodium sulfate. The reaction products of VX are CO_2, H_2O, nitrogen, sodium phosphate, and sodium sulfate. The CO_2, H_2O, excess O_2, nitrogen, and some vaporized salts exit the reactor as offgas with most of the salts remaining in the reactor.

The residual waste generated by the molten salt oxidation process is spent salt that flows out of the reactor into 55-gallon steel drums where it solidifies. There are no aqueous effluents from the process.

The process can treat both HD and VX liquids that have been drained from chemical weapons or bulk storage. The process has not been conceptually designed to treat metal parts or drained ton containers but can treat decontamination liquids. Organic destruction efficiencies have exceeded 99.9999 percent (six nines) in some cases. Results indicate that soot, char, and chlorine accumulations in the melt need to be controlled to assure efficient molten salt oxidation operation. Depending on the waste to be treated and the operating temperature, a high-temperature, corrosion-resistant alloy or ceramic-lined vessel may be used to contain the melt. A suitable offgas system is required to remove particulate, volatilized metals, and any products of incomplete combustion.

The molten salt oxidation process can also be operated with oxygen rather than air to minimize offgas production during agent destruction. Operation with oxygen will allow the reduced quantity of offgas to be captured and certified clean before it is ultimately released.

Molten salt oxidation was developed by Rockwell International in the 1950s to process nuclear fuels. Waste treatment applications were recognized and bench-scale, pilot-scale, and demonstration-scale units with capacities up to 900 kg/h (2,000 lbs./h) were subsequently conducted in which a variety of organics and chemical weapons agents were destroyed (Stelman et al., 1992; Gray, 1991; and U.S. Department of Energy, 1993). A DOE peer review (Cudahy et al., 1993) identified a number of unresolved molten salt oxidation issues (for example, the pretreatment of solids, materials of construction, melt freeze-up, buildup and monitoring of residues in the melt, and molten salt recycle). Some of the issues have not been fully resolved. DOE's Lawrence Livermore National Laboratory (LLNL) has tested a two-stage bench-scale molten salt oxidation unit (Crosley et al., 1995) and reportedly has a larger single-stage unit in operation.

9. Alkali Metal Reaction

A novel method for disposing of toxic organic chemicals uses reduction reactions between organic and metallic sodium resulting in solid products (Chan, 1993). The advantage of the method is that the emission of gaseous products is controlled by process parameters and can be adjusted for minimal gaseous products or a pure hydrogen or hydrogen and methane stream.

The process depends on the submerged reaction of the organic vapor in liquid sodium operating in the temperature range from 752° to 932°F (400° to 500°C). The reaction process is controlled through the use of a choked-flow sonic or critical flow nozzle. With the injection nozzle operating at critical flow conditions, the feed rates are independent from downstream pressure variations (within limits). The choked-flow nozzle serves two purposes: 1) it isolates the vapor generator from the liquid metal system, protecting against the possibility of liquid metal from backing up into the organic feed system and causing potentially dangerous contact with the liquid organic feedstock, and 2) it delivers the vapor at a fixed rate, independent of the temperature and pressure fluctuations in the reaction zone providing for control of reaction kinetics.

Laboratory tests using 4 grams of 1,1,1-trichloroethane were conducted at ANL in

October 1994, to investigate alkali metal reactions. The reductive reactions were rapid and efficient and no trace of the feed compound was detected in the effluent. No chlorine, either as any gaseous compound or elemental chlorine, was detected in the product gas. It is assumed that chlorine formed an alkaline salt and remained in the batch reaction chamber.

Data for this technology are insufficient to support an estimate of potential process effluents.

10. Gas-Phase Reduction Process

The Eco Logic process is based on the high-temperature gas-phase reduction of chemical warfare agents in an atmosphere of steam and excess hydrogen (Holm et al., 1994) (see *Figure* 6 and Tables 6a and 6c). The reduced chemical agents are converted to simpler molecules (CO, CO_2, HCl, NH_3, PH_3, and H_2S) and a solid residue from metal or minerals that are introduced, with the feed, into the process. A large fraction of the product gas is recycled to the reactor to take advantage of the high concentration of hydrogen present. Gas removed from the process passes through a multiple-stage scrubber system to remove H_2O, mineral acids, and CO_2. The remaining "clean" synthesis gas is then compressed to about 100 pounds per square inch gauge (psig) and stored in tanks until it can be tested and certified to be clean. The synthesis gas is used to offset fuel requirements for raising steam and heating a methane reformer. It is expensive but possible to recover mineral acid effluents for recycle. The mineral acids could also be neutralized and the solid residues dried for landfill disposal.

Decontamination of drained ton containers would be accomplished in an indirectly heated chamber using a hydrogen sweep gas until the containers reach a 5X level of decontamination. The sweep gas along with other evaporated and pyrolyzed offgases are fed into the main process reactor. Dunnage, metal parts, spent carbon, and other solids can be treated in the same manner.

Liquid agent drained from the ton containers is pumped to the main process vessel and sprayed into the high-temperature [1,562°F (850°C)] gas-phase environment. The agent spray evaporates and reacts without a catalyst at this temperature. Thermal reduction of the agent is highly endothermic. Electric radiant heaters located inside the primary reactor provide energy. The reactor design is similar in geometry to a cyclone. The agent is sprayed into the reactor and mixes into the swirling flow around the radiant heaters. At the bottom of the reactor, a large-diameter downcomer centered on the reactor axis conveys the reacted gas up and out through the top of the reactor. Because of the need to prevent oxygen from entering the reactor, it operates at a positive pressure (10-inches water column) to prevent the infiltration of air.

The tempered product gas enters the condenser, scrubber, and polishing train. In the condenser, HCl and water vapor are removed along with some NH_3 and phosphine (PH_3). Three scrubbers are used: an 80 percent sulfuric acid (H_2SO_4) scrubber, a caustic scrubber, and a monoethanolamine (MEA) scrubber. Acid gases are removed in the first two scrubbers and the third scrubber will remove the majority of CO_2 and any condensable hydrocarbons remaining in the gas.

Following the scrubber train, the product gas is compressed and stored in gas cylinders. This permits sampling and analysis of the gas to ensure that it is free of agent and any toxic byproducts of agent destruction. The gas, once certified, is then used to offset auxiliary fuel consumption in the process unit operations.

Eco Logic has indicated that their facility is mobile and the same equipment could be used at Aberdeen to treat HD, then decontaminated and moved to Newport, where VX would be destroyed.

11. Hydrotreating

Catalytic hydrotreating is widely used in the petroleum refining industry. It is primarily used to remove compounds containing oxygen, nitrogen, or sulfur from various petroleum fractions by converting them into water (H_2O), NH_3, H_2S, and hydrocarbons. Hydrotreating technology for the catalytic destruction of toxic chlorinated hydrocarbons has been proposed as an alternative to incineration. In treating chemical warfare agents, hydrogen could be used to capture chlorine (Cl), phosphorus (P), and sulfur (S). Hydrotreating processes operate in a temperature range from 752° to 932°F (400°to 500°C); however, pressure ranges vary widely from 10 to 100 atmospheres (150 to 1,500 psia). Excess hydrogen is needed (80 volume percent) and, depending on the reaction desired, a suitable catalyst is required.

Conceptually, hydrotreating could be used to decontaminate metal parts using an auxiliary indirectly heated chamber blanketed with a hydrogen sweep gas. Operating temperatures would need to be at least 1,000°F (537°C) to achieve the military standard of 5X decontamination of metal parts. This concept has not been demonstrated.

Hydrotreating would not be expected to succeed for propellants or other energetic materials that contain their own oxidizers. Process temperatures would have to be controlled below autoignition to prevent uncontrolled oxidation and the resulting temperature or pressure excursions.

The expected offgas from treating chemical warfare agents would contain acid gases and a fuel or synthesis gas. After reducing the offgas temperature, acid gases could be neutralized and scrubbed to form a wastewater brine. Alternatively, there is also the potential for acid recovery from the offgas. The synthesis gas would be captured and probably compressed, while analytical gas certification procedures were used to characterize it as a byproduct. A solid waste stream composed of spent catalyst would be taken from the catalytic hydrotreating reactor.

12. Supercritical Water Oxidation

SCWO reacts combustible materials using air or oxygen in water at temperatures and pressures above the critical point of water, 705°F (374°C) and 218 atmospheres [3,205 pounds per square inch (psi)] (see *Figure* 7 and Tables 7a through 7c). Oxidation is accomplished under single-phase (liquid) conditions to obtain excellent dense-phase mixing and high rates of heat transfer.

SCWO uses a two-step process for the treatment of bulk agent: 1) neutralization of agent by chemical means developed by the Army under its ongoing program, and 2) treatment of the neutralized agent by SCWO technology for mineralization of the nonaqueous constituents. The process cannot treat liquid agent directly since it is necessary to dilute the agent to control the highly exothermic reactions as it is being pumped into the reactor.

The SCWO process does not have the ability to treat ton containers or other solids, such as dunnage, waste activated carbon, or discarded personal protection equipment (PPE). SCWO technology does, however, have the ability to effectively treat the aqueous decontamination liquids coming from alternative processes used for treating solid materials.

The performance of the SCWO process to date warrants consideration as a posttreatment method (instead of, or in conjunction with, biological treatment) for the product stream from the chemical neutralization process. This reflects recommendations made by the NRC in their February 1994 report on alternative technologies.

13. Wet Air Oxidation

WAO is an aqueous-phase oxidation process using molecular oxygen contained in air as an oxidant (Holm et al., 1996) (see *Figure* 8 and Tables 8a through 8c). The process operates at elevated temperatures ranging from 390° to 625°F (200° to 300°C) and pressures up to 200 atmospheres (3,000 psi).

WAO would require the following three-step process for the treatment of bulk agent: 1) neutralization of agent by chemical means developed by the Army under its ongoing program, 2) treatment of the neutralized agent by WAO technology, and 3) liquid effluent discharge to a biological treatment plant for final polishing. The process cannot treat liquid agent directly since it is necessary to dilute the agent to control the highly exothermic reactions as it is being pumped into the reactor.

The WAO process does not have the ability to treat ton containers or other solids, such as dunnage, waste-activated carbon, or discarded PPE. WAO technology does, however, have the ability to effectively treat the aqueous decontamination streams produced by processes currently under investigation for treating these solid materials.

The performance of the WAO process to date warrants consideration as a posttreatment method (in conjunction with biological treatment) for the product stream from the chemical neutralization process. This reflects recommendations made by the NRC in their February 1994 report on alternative technologies.

14. Steam Reforming

Steam reforming can be used to destroy chemical warfare agents by reacting them with superheated steam (see *Figure* 9 and Tables 9a through 9c). A synthesis gas consisting of CO, CO_2, H_2, H_2O, and CH_4 is produced from these wastes. Steam reforming is mainly a two-stage operation, one stage where organics are vaporized by exposure to 600°F (316°C) to 1,100°F (593°C) superheated steam, and a second stage where the first stage offgas is mixed with 2,000°F (1,093°C), or greater, superheated steam to complete the reforming reactions. Steam reforming is performed at atmospheric pressure, significantly reduces waste volume, and takes place in a low oxygen-reducing environment to avoid production of dioxins and furans.

Catalytic steam reforming is a common industrial process that is used to make hydrogen gas from methane. Noncatalytic steam reforming has been studied as a means of gasification of coal and cellulosic wastes. Sandia National Laboratory and LLNL, both in Livermore, California, and three vendors, Synthetica Technologies, Inc., ThermoChem, Inc., and Plasma Energy Applied Technology, Inc., are active in steam reforming. Bench-scale and pilot-scale tests have been conducted (Aghamohammadi et al., 1993 and Galloway et al., 1993). Results show that two-stage organic destruction efficiencies range between 99.99 and 99.999 percent (four nines and five nines, respectively). With the addition of air emission abatement devices, only benzene has been detected, at very low levels, in steam reforming airborne emissions.

Synthetica Technologies, Inc., has obtained a regulatory opinion from the State of California and the district EPA office that their system is not an incinerator and would be considered a RCRA unit permitted under Subpart X. However, HCl production from high-chlorine content waste remains a problem and activities related to removal of chlorine from the first stage offgas are in progress. This particular process is energy intensive and process heat must be removed by suitable cooling methods.

ThermoChem, Inc., offers a sodium carbonate fluidized bed system (Aghamohammadi, 1995) that is a fairly mature process for biomass waste treatment. A 65 tonne/day (72 ton/day) unit has been installed in North Carolina, and a 45 tonne/day (50 ton/day) unit became operational in India in 1993. Tests on DOE mixed waste surrogates have recently been completed (Bostick et al., 1994). A second stage, fixed catalytic reactor, is under development. Low operating temperatures may prevent complete destruction of many hazardous organics. Higher operating temperatures, however, may cause melting of some wastes and plugging of the bed. The waste to be treated and the addition of the second stage reformer could alleviate these concerns.

Plasma Energy Applied Technology, Inc., Huntsville, Alabama, has developed a single-stage, plasma torch, Thermal Destruction and Recovery (TDR) system for destroying organic waste in a steam reforming mode. Steam is injected into the reactor vessel to reform the gases into useful byproducts. Gaseous products are similar to other steam reforming processes. An "Authority to Construct" has been issued by the San Diego County Air Pollution Control District to install a TDR system for processing medical waste at Kaiser Permanente Medical Center in San Diego, California. This could be the first permitting of a plasma waste processor in the United States.

15. Catalytic Oxidation

Thermal catalytic oxidation (TCO) has developed into an abatement method for chlorinated volatile organic compounds (VOCs) (Holm et al., 1994) (see *Figure* 10 and Tables 10a through 10c). This catalysis process primarily converts gaseous organic waste into CO_2 and water. Organic destruction occurs at temperatures up to 1,112°F (600°C), energy requirements are less than those for incineration, and an offgas treatment is required to treat entrained products of incomplete combustion and acidic gases. Early applications of this process required frequent recycling or regeneration of the catalyst. Current catalysts are improved but still subject to process interactions (for example, sulfur and phosphorus poisoning, metal sintering, masking, and fouling) that reduce the available active metal surface area.

A TCO field test demonstration has been conducted at the Savannah River Site, Aiken, South Carolina. The demonstration was conducted as part of DOE's Office of Technology Development VOCs in Non-Arid Soil Integrated Demonstration. A Johnson-Matthey halocarbon destruction catalyst was used in the demonstration. Results show that organic destruction efficiencies ranged up to 99.99 percent (four nines) for perchloroethylene and up to 99.9 percent (three nines) for trichlorethylene. Results also show that destruction efficiencies decreased dramatically at temperatures below 932°F (500°C) (Jarosch et al., 1994).

Catalytic oxidation at temperatures in the 662° to 932°F (350° to 500°C) range are being proposed by Allied Signal. They have recommended that their catalytic oxidation system be used in conjunction with the baseline incineration systems as a safety enhancement. Allied Signal has considered using catalytic oxidation in three ways: as an improved secondary combustion process, as a polishing step after secondary combustion but before the pollution abatement system (PAS), and as a polishing step after the PAS. The use of catalytic oxidation as a replacement for carbon filter systems has significant benefits that may warrant further evaluation.

16. Reaction with Sulfur

The sulfur process reacts organic compounds with elemental sulfur, either in a vapor phase or liquid phase at temperatures ranging from 275° to 850°F (135° to 450°C) (see *Figure* 11 and Tables 11a through 11c). Higher temperature may be present in the vapor phase. Using a nitrogen sweep gas, waste materials are combined with sulfur and fed into a two-stage reactor. For chemical warfare agents, reactions in molten sulfur or sulfur vapors are analogous to oxidative reactions; however, the reactions occur at lower temperatures and produce different byproducts. Example byproducts from the sulfur reactions include H_2S, SO_2, carbon disulfide (CS_2), HCl, nitrogen oxides (NO_x, depending on the presence of oxygen), phosphorus sulfides, sulfur chlorides and oxychloride, and a polymeric form of carbon-sulfur.

The gas-phase reactants will be cooled and passed through a cyclone to remove entrained solids. The gases will then flow into a wet pollution abatement process for acid gas removal and to heat exchangers to control condensable offgases.

Although the sulfur process could be used to decontaminate metal parts, no conceptual design plan has been developed for this purpose. Operating temperatures would have to be higher than the proposed 850°F (454°C) to achieve the military standard 5X decontamination of metal surfaces. Experimental work has been conducted for sulfur vapor-phase reactions in the temperature range from 1,000° to 3,000°F (537° to 1,648°C) on five chlorinated compounds; however, the test results did not indicate complete destruction of these materials (University of Pittsburgh Applied Research Center, 1991).

Tests on the Adam's Sulfur Process have been performed at Picatinny in 1993 (Chen et al., 1993) to investigate the destruction of energetics through reactions with sulfur. Results, based on 500 milligram tests on TNT, RDX, HMX, and PETN, indicate that the reaction rate is a strong function of temperature and must be controlled below autoignition temperature. In cases where the energetic material contains its own oxidizer, autoignition would drive a rapid oxidation reaction that would approach adiabatic combustion temperatures.

Experimentation on the Adam's Process at Pine Bluff Arsenal (Wright, 1995) has demonstrated that reactions take place between hot sulfur gases and dilute HD. The reactants were mixed in a stainless steel reactor and heated to 857°F (460°C). The evidence that these reactions took place was collected by examination of their reaction products using gas chromatography (GC)/mass spectroscopy (MS). The evidence indicates that the Adam's Process is effective in destroying dilute mustard agent and simulant agents; however, more experiments are necessary to determine destruction efficiencies, kinetics, and to provide scientifically valid conclusions.

17. Thermal Media Oxidation

The packed-bed reactor is a flameless, thermal oxidizer that destroys vapor-phase organics (Schwindendorf et al., 1995 and Holm et al., 1996). The inert, noncatalytic-packed bed consists of an insulated cylinder lined with alumina and silica refractory that contains a ceramic metal oxide matrix. The bed is operated at 1,600°F (871°C) to 1,850°F (1,010°C), being first heated by an external preheater and sustained at operating temperature by the heat of oxidation of the organic vapors. The vapors are combined with oxygen in the mixing zone of the bed and rise into the reaction zone, where the gas temperature increases rapidly and the organics are oxidized. Passage of the gas through the small interstitial spaces within the matrix generates a high degree of turbulence. This results in efficient mixing of oxygen with the organics in the mixing zone and effective heat transfer between the gases and the high temperature reaction zone. Oxidation is usually completed within the first few inches of the ceramic matrix reaction zone, allowing for a relatively compact system.

Between 1987 and 1991, three experimental units were built by Thermatrix, Inc., for internal development purposes. These units were used to verify destruction efficiency, emission control, and operational characteristics (Binder et al., 1993 and Thermatrix, 1992). Twelve units have been installed or are under construction (Woods et al., 1993).

Thermatrix has designed a system to process offgas from a thermal desorber and is developing an integrated thermal desorption system capable of treating soils contaminated with volatile organics (Wilbourn et al., 1994).

Thermal media oxidation at temperatures in the 1,652° to 1,832°F (900° to 1,100°C) range are being proposed by Thermatrix, Inc. They have recommended that their packed-bed oxidation system be used in conjunction with the baseline incineration systems as a safety enhancement. Thermatrix, Inc., has considered using their oxidation process in three ways: as an improved secondary combustion process, as a polishing step after secondary combustion but before the PAS, and as a polishing step after the PAS. The use of thermal media oxidation as a replacement for carbon filter systems may have significant benefits that may warrant further evaluation.

18. Gasification Process

AGGCOM is a two-stage operation combining two coal gasification technologies (U.S. Environmental Protection Agency, 1991 and Mensinger et al., 1991) (see *Figure* 12 and Tables 12a through 12c). The first stage is based on a sloping grid, fluidized-bed technology, which was developed as a part of a coal gasification process. The second stage is a cyclonic converter. The bulk of the fluidized bed operates at a temperature of 1,500° to 2,000°F (816° to 1,093°C), while the central spout from the bed operates at a sufficiently higher temperature to partially fuse and agglomerate the ash. The mobility of inorganic contaminants, such as metals, in moving into the converter is reduced because of their incorporation into the glassy matrix agglomerate from the SGFB. Additional destruction of gaseous products leaving the first stage is achieved in a cyclonic combustor.

An 0.91 m (3 foot) diameter unit employing the sloping grid technology was tested with coal (over 10,000 hours of operation) and demonstrated that agglomerated ash can be produced. These agglomerates were tested by the Extraction Procedure Toxicity Test and found not to be hazardous waste under the applicable definition. The same unit was tested with spent blast abrasive feeds and a spent foundry sand at rates of approximately 454 kg/h (1,000 lbs./h) to 907 kg/h (2,000 lbs./h). The wastes were contaminated with 1 to 2 percent organics. In the test with the spent blast abrasive feed, it was determined that the organic destruction exceeded 99.99 percent for the tributyl tin oxide contaminant and the reclaimed blast abrasive was suitable for reuse under U.S. Navy specifications. In separate testing, the cyclonic combustor has been evaluated in an 0.84 Gj/h (0.8 MM Btu/h) unit at a feed rate of 13.6 kg/h (30 lbs./h) to 27.2 kg/h (60 lbs./h) of synthetic waste and achieved over 99.9999 percent (six nines) destruction efficiency for CCl_4.

More recently, a series of soil agglomeration tests were conducted with raw and spiked samples of soil in a 15.2 cm (6 inch) diameter fluidized-bed bench-scale unit. Agglomerated or vitrified soil samples were produced in four of the eight tests. Operating conditions were confirmed, samples passed the TCLP tests, and results indicate that higher contaminant levels can be processed by AGGCOM (Rehmat et al., 1995).

This technology was accepted into the Superfund Innovative Technology Evaluation (SITE) Emerging Technologies Program in July 1990. A 5.4 tonne/day (6 tons/day) two-stage, soil pilot plant has been constructed and tested, producing additional agglomerated soil samples under various operating conditions. The second-stage cyclonic combustor in this plant was effective in destroying carbon tetrachloride [a surrogate for polychlorinated biphenyls (PCBs)]. Future testing of the AGGCOM process will focus on the sustained and continuous operation of the pilot plant with soil admixed with both organic and inorganic surrogate compounds. Plans are to test other feed stocks, such as industrial waste, automobile shredder residue (fluff), and medical wastes.

19. Solvated Electron Technology

"Solvated electron solutions" are powerful reducing agents, capable of reacting with agent and other hazardous organic compounds to produce less hazardous organic species and inorganic salts to be disposed of in an environmentally safe fashion. Dissolving alkali or alkaline earth metals (for example, sodium, potassium, lithium, and calcium) produces these solutions in anhydrous liquid ammonia. Controlling the rate of addition of metal reactant and agent controls the reaction. Although this chemistry has been known since the mid 1800s, commercial applications are few. Formation of the reactive solution occurs via equation 1.

$$Na + NH_3 \rightarrow Na^+ + e^- \qquad (1$$

Destruction of a chlorinated organic is stated to take place through the general reaction given in equation 2.

$$C_{12}H_4Cl_6 + 6Ca + 6H_2O \rightarrow C_{12}H_{10} + 3Ca(OH)_2 + 3CaCl \qquad (2$$

The technology has been used commercially for chlorofluorocarbons (CFCs) and PCBs and has been successful in destroying chemical warfare agents. PCBs and CFCs, with which it has been commercially demonstrated, have substantially different physical and chemical properties from agent. Destruction removal efficiencies (DREs) of about 99.99 percent have been achieved with these compounds, but no six-nines level has been demonstrated with any compound.

Data for this technology are insufficient to support an estimate of potential process effluents.

20. Caustic Hydrolysis

The process for destruction of VX at Newport, Indiana, is neutralization followed by shipment to an offsite treatment, storage, and disposal facility (TSDF) for biodegradation (U.S. Army Concept Design Package for VX, May 1996). This process (see *Figure* 13 and Table 13a) involves neutralization of VX with sodium hydroxide at 194°F (90°C). At 33 percent VX and 67 percent caustic (20.6 wt % sodium hydroxide), the reaction produces a solution containing sodium o-ethyl methylphosphonate, the corresponding thiolamine, and EA 2192 (parts per million levels). This product is approximately 40,000 times less toxic than the original VX. After cooling, the mixture is reacted at room temperature with one volume of 5 percent sodium hypochlorite. After testing to ensure agent destruction, the effluent is released from the Toxic Cubicle and further treated at room temperature with water and sodium hypochlorite. The resulting material is then tested again to ensure that VX is below detection limits of 20 parts per billion (ppb) and held for shipment to a commercial hazardous wastewater biotreatment facility.

The ton containers are drained, flushed, cut in half, and cleaned with hot, high-pressure water and steam. The resulting effluent is used in the neutralization process, replacing part of the required water. The exact ratio will depend on VX and organic carbon content. The ton containers are monitored for 3X condition and then sent to Rock Island Arsenal (RIA), Illinois, for 5X treatment and smelting. The vapors from the reactor are scrubbed with caustic, cooled to remove water vapor, reheated sufficiently above the saturation temperature, filtered in carbon filter beds, and then released through the plant ventilation system.

A detailed report, entitled Newport Chemical Activity Concept Design Package for VX Neutralization followed by Offsite Biodegradation, was submitted to the NRC by PMCD, Aberdeen Proving Ground (APG), Maryland, in May 1996. Information on process concepts and material balances are included in that report.

21. Water Hydrolysis

For bulk-loaded HD, stored at APG, neutralization followed by biodegradation has been investigated as the neutralization-based option by the Army (U.S. Army Concept Design Package for HD, May 1996) (see *Figure* 14 and Table 14). The NRC has developed a design based on “no-liquid” discharge for review. The design can accommodate neutralization and offsite shipment to a commercial wastewater treatment facility or onsite biodegradation and discharge of the liquid effluent into the APG wastewater treatment plant. The "zero-liquid discharge" option is considered the most conservative option for APG.

HD is initially neutralized by vigorous mixing in water at 194°F (90°C) using a 4 percent feed of agent. The reaction is 90 percent selective to thiodiglycol (TDG) and HCl. Once the reaction is complete, caustic is added to adjust the pH to 12. After testing to ensure agent destruction (less than 200 ppb HD; limit of detection is 12 ppb), the hydrolysate is released from the Toxic Cubicle and fed to a falling film evaporator, where the volatile organic carbon compounds are stripped out, condensed, and combined with the water recycle stream for subsequent destruction in a UV oxidation unit. The remaining hydrolysate is fed to sequencing batch bioreactors (SBRs) (see *Figure* 15 and Table 15).

The SBRs consume more than 90 percent of the organic carbon, including 99 percent of the TDG. This level of destruction meets the Chemical Weapons Convention (CWC) definition for destruction of TDG as a schedule 2 compound. The resulting liquid effluent is filtered and recycled through an evaporator. The evaporator bottoms consist of salt brines, which are mixed with solidifiers and disposed of in a landfill. The waste biomass from the SBRs is fed to an aerobic digester to reduce the volume, then dried in a filter press, and disposed of commercially. Maryland law requires that these solid wastes be treated as hazardous wastes, even though they have no hazardous constituents. Approximately 6 pounds of solid waste is produced for each pound of HD destroyed.

The ton containers are drained, flushed, cut in half, and cleaned with hot, high-pressure water and steam. The resulting effluent is used in the neutralization process, in place of part of the required process water for HD. The exact ratio will depend on the HD and organic carbon content. The ton containers are monitored for 3X condition [less than time-weighted average (TWA)] and then sent to RIA to be smelted. The vapors from the reactor and hydrolysate tanks are scrubbed with caustic, cooled, superheated, filtered in guard beds, and then released through the plant ventilation system. The gas stream from the SRBs is filtered for odor control and released to the atmosphere.

A detailed report, entitled Aberdeen Proving Ground Concept Design Package for VX Neutralization followed by Offsite Biodegradation, was submitted to the NCR by PMCD at APG in May 1996. Information on process concepts and material balances are included in that report.

22. Ultraviolet Irradiation and Oxidation

Oxidation with UV light and oxidizing agents can be used for destruction of hazardous compounds (Engleman and Holm, 1995). The processes are sometimes called photochemical degradation processes or Advanced Oxidation Processes (AOP). There are several types of AOPs, including H_2O_2/UV, ozone/UV, ozone/H_2O_2/UV, photocatalyzed UV, and vacuum UV processes.

The oxidation of hazardous organic compounds, using UV light and oxidants, involves the production and subsequent reaction of hydroxyl radicals. The hydroxyl radical is a powerful, but short-lived, oxidizing agent. Production of the hydroxyl radical in situ can lead to destruction of the contaminant compounds. The first three processes (H_2O_2/UV, ozone/UV, and ozone/H_2O_2/UV) use the UV photolysis of hydrogen peroxide and/or ozone to generate hydroxyl radicals. Photocatalyzed UV technology uses a catalyst (such as a semiconductor or other catalyst) to generate the hydroxide radical from water and hydroxide ions. The vacuum UV process uses high-energy UV radiation to generate hydroxyl and hydrogen radicals from water.

In addition, solar UV processes are being developed that can destroy hazardous compounds in soils or as higher concentration chemical waste. There are two different approaches under development, photolytic processes and catalytic steam reforming. Photolytic processes use the visible and infrared portions of the solar spectrum to desorb volatile molecules from soil or vaporize a chemical waste and then enhance the destruction of target molecules using the near-UV solar radiation. Catalytic steam reforming uses a rhodium catalyst to absorb the solar radiation and assist the reaction of steam and target molecules.

22.1. H_2O_2/UV PROCESS

UV photolysis of hydrogen peroxide leads to two hydroxyl radicals per quantum of radiation absorbed.

hν

$H_2O_2 \rightarrow 2HO\bullet$

The rate of photolysis is dependent on the pH and is higher under alkaline conditions. The hydroxyl radicals oxidize organic pollutants dissolved or dispersed in aqueous systems. Sundstrom et al. investigated the efficiency of the H_2O_2/UV process with a variety of aliphatic and aromatic compounds, including chlorinated compounds. They found that the reactivities of volatile aromatic halocarbons fell in the order of polychlorinated ethylene > trichloroethylene > chloroform > methylene chloride > tetrachloroethane, ethylene dibromide, and carbon tetrachloride. The reactivities of aromatic compounds were found to be in the order of trichlorophenol > toluene > benzene > dichlorophenol > phenol > chlorobenzene >chlorophenol >diethyl phthalate and dimethyl phthalate. The chlorine substituents from the compounds were converted to chloride ions, indicating that the chlorinated structures were destroyed.

Data for this technology are insufficient to support an estimate of potential process effluents.

23. Electrochemical Oxidation With Silver

The Silver 2 electrochemical oxidation process operates below 212°F (100°C) and ambient pressure using a nitric acid reaction medium (see *Figure* 16 and Tables 16a and 16b). The process relies on the use of an electrically charged silver catalyst to generate highly reactive conditions for organic materials oxidizing them to CO_2, water, and inorganic compounds. This technology has been used for the destruction of a wide variety of organic components found in radioactive waste. The process uses DC electrical power distributed to electrochemical cells, each equipped with one anode and one cathode. Example byproducts from this process include a gas stream containing N_2, H_2, NO_x, CO_2, CO, a brine containing metallic salts, and waste acid.

Decontamination of drained ton containers would be accomplished inside a double-walled vessel. The container would be connected to the electrochemical reactor using special flanges and triple-walled transfer lines. The analyte would circulate through the ton container until a 3X level decontamination is achieved. For process control and agent detection, gas vent, gas headspace, and liquid volumes would be carefully monitored.

As chemical warfare agents are introduced, the electrochemical process oxidizes the carbon, nitrogen, sulfur, and phosphorus in the agent to form products, including CO_2, CO, NO_x, sulfates, phosphates, and phosphorus pentoxide. Hydrogen in the agent is converted to water and chloride is converted to salts. Oxidation is primarily directed at the conversion of carbon to CO and CO_2. The oxidation occurs through the use of silver ions in the plus 2 oxidation state, Ag^{2+}. Silver ions gain electrons from interactions with carbon (or N, P, and S) and return to Ag^+ in a thermodynamically favorable reaction. Oxidation of the products eliminates any reverse reaction or reformation of agent. Ag^{2+} is recovered at the anode from silver nitrate ($AgNO_3$), and nitric acid (HNO_3) is reduced at the cathode.

Several secondary treatment units are used in the Silver 2 process: NO_x reformer, silver nitrate recovery, silver management, anolyte offgas condenser, and offgas treatment. An NO_x reformer is used to process NO_x and nitrous acid and recovers nitric acid for the catholyte circuit. The acid reformer includes a boiler/preheater and oxygen addition. The reformer boiler is also used to obtain a concentrated mixture of silver nitrate and nitric acid. The mixture is separated in a crystallizer and the silver nitrate is recycled to the anolyte circuit. Insoluble silver chloride is removed continuously from the system by a cyclone separator for recovering the silver.

The offgas from the electrolytic cells and the nitric acid reformer is a mixture of CO_2, water vapor, oxygen, nitrogen and NO_x. Condensable vapors in the offgas are removed and hydrogen peroxide scrubbers are used to control NO_x and VOCs. An activated carbon filter is used as the final pollution abatement step.

AEA Technology has proposed first operating their electrochemical oxidation facility at Aberdeen, treating HD. On completion of stockpile destruction at Aberdeen, the equipment would be decontaminated and moved to Newport, where VX would be destroyed.

24. Electrochemical Oxidation With Cerium

The cerium-mediated electrochemical oxidation (MEO) uses electricity to drive an aqueous-based chemistry that destroys organic hazardous organic substances in a regenerative system (Holm et al, January 1996). The process uses the anodic activation of the metal ion mediator in an electrolyte to stimulate oxidative reactions with toxic organic compounds. The cerium mediator, which becomes the reactive metal ion, is recycled through reoxidation at the anode.

The primary reaction products from the reaction of the organic materials are CO_2 and H_2O. Heteroatoms contained in the organic materials are also oxidized; sulfur-to-sulfate, phosphorus-to-phosphate, and carbon bound chlorine-to-HCl. Bound fluorine is released as fluoride ion if the MEO process operates below the fluorine-fluoride potential (0.6 to 0.8 volt).

The MEO process operates near room temperature and at atmospheric pressure in an aqueous electrolyte containing a mineral acid. The choice of acid, primarily nitric acid, depends on the metal ion and the heteroatom composition of the waste stream. The process achieves destruction of organic material due to the thermodynamic and kinetic activity of the metal ion mediator. The highly oxidative environment is achieved by direct application of substantial amounts of electrical power to drive an endothermic reaction, rather than the high-temperature exothermic and autogenous oxidation produced by combustion.

Decomposition of the waste organic materials occurs in the anolyte (anode electrolyte) where the oxidized metal ion mediator is produced. The destruction process can be heterogeneous as the reaction medium often consists of a finely dispersed emulsion of the organic phase in the aqueous electrolyte, the organic oxidation occurring at the electrolyte-organic phase interface.

Data for this technology are insufficient to support an estimate of potential process effluents.

25. Hydrogenation (Base Catalyst Decomposition)

The Base Catalyst Decomposition (BCD) process is catalytic hydrogenation operating in a 212° to 662°F (100° to 350°C) temperature range (Holm, 1994). The major departure from other hydrogenation processes is an attempt to maintain a low-pressure and low-temperature reactor. Wright State University has 1-liter bench-top reactors capable of low- and high-pressure batch operation. The proposed hydrogenation process would mix sodium hydroxide agent and hydrogen with a catalyst detoxifying the mixture, primarily by caustic hydrolysis at elevated temperature/pressure, and drive the reactions to an irreversible state through the reactions that occur with hydrogen. Several hydrogen donors were discussed (fuel oil, no. 6 oil, etc.), but no specific recommendation was suggested. Although the NRC Alternative Technology Committee briefly discussed this technology, it was not specifically analyzed by the committee or stated in their report (Longwell et al., 1993).

A broad range of halogenated hazardous and toxic wastes were tested at Wright State University using BCD: chlorophenols, PCBs, polychlorinated dioxins, polychlorinated dibenzofurans, chlorophenoxy herbicides, DDT, Lindade, Toxophane, and other pesticides. Liquid wastes can be treated directly; however, contaminated soils would require thermal desorption to separate and concentrate the liquid contaminants for treatment in the BCD reactor. The Smith's Farm Superfund cleanup of PCBs had planned to use BCD technology to treat condensed vapors from thermal desorption, but the small amounts of contaminated liquid did not warrant use of the BCD reactor. Instead, the BCD chemical mixtures were combined with the contaminated liquid stream and recycled through the thermal desorption facility.

Data for this technology are insufficient to support an estimate of potential process effluents.

Figure 2. Molten Metal Technology

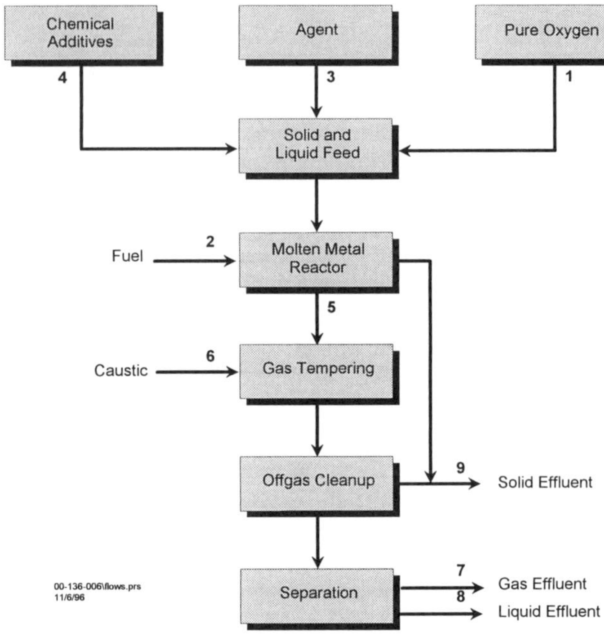

TABLE 2a. Molten Metal Technology (per unit mass of chemical agent VX)

STREAM	1	2	3	4	5	6	7	8	9
Flow Components	Oxygen	Fuel	Chemical Agent		Gas		Gas Effluent	Liquid Effluent	Solid Effluent
C		0.02	0.49						
H		0.01	0.10						
O	0.54		0.12						
P			0.12						0.12
S			0.12						0.12
N			0.05		0.05		0.05		
H_2O									
CO					1.45		1.53		
H_2					0.13		0.14		
Other								1.23	
TOTAL[a]	0.54	0.03	1.0		1.62		1.73	1.23	0.24

Total Input=3.19 Total Output=3.19

a Round-off error causes some of the totals to be 0.1 lb higher/lower than the sum of all the constituents in some columns.

TABLE 2b. Molten Metal Technology (per unit mass of chemical agent GB)

STREAM	1	2	3	4	5	6	7	8	9
Flow Components	Oxygen	Fuel	Chemical Agent	Lime	Gas	Caustic Quench	Gas Effluent	Liquid Effluent	Solid Effluent
Methane		0.0							
Agent			1.0						
N_2				0.0	0.0		0.0		
O_2	0.5								
H_2					0.1		0.1		
H_2O						1.4		1.4	
CO					0.9		0.9		
NaOH						0.5		0.4	
CaO				0.9					0.4
CaF_2									0.1
NaF								0.2	
$Ca_2P_2O_7$									0.9
HF					0.1				
Other[a]				0.1			0.0		
TOTAL[b]	0.5	0.0	1.0	1.0	1.0	1.9	1.0	2.0	1.4

Total Input=4.4 Total Output =4.4

a This technology is not yet fully understood, so unknown products may have been involved that contributed carbon and phosphorus to the system and unknown reactants may have been involved that contributed calcium and nitrogen.

b Round-off error causes some of the totals to be 0.1 lb higher/lower than the sum of all the constituents in some columns.

TABLE 2c. Molten Metal Technology (per unit mass of chemical agent HD)

STREAM	1	2	3	4	5	6	7	8	9
Flow Component	O_2	Fuel	Chemical Agent	Other	Water	Caustic Quench	Gas Effluent	Liquid Effluent	Solid Effluent
C		0.02	0.31			0.12			
H		0.01	0.05			0.02			
Cl			0.44						
S			0.21						0.21
O	0.53								
H_2O					0.99			0.99	
CO							1.12		
H_2							0.07		
HCl								0.47	
Other[a]				0.16					
TOTAL[b]	0.53	0.03	1.0	0.16	0.99	0.14	1.19	1.46	0.21

Total Input=2.85 Total Output=2.85

a This technology is not yet fully understood, so unknown reactants may have been involved that contributed carbon, chlorine, and hydrogen to the system.

b Round-off error causes some of the totals to be 0.1 lb higher/lower than the sum of all the constituents in some columns.

Figure 3. Carbon Electrode Technology

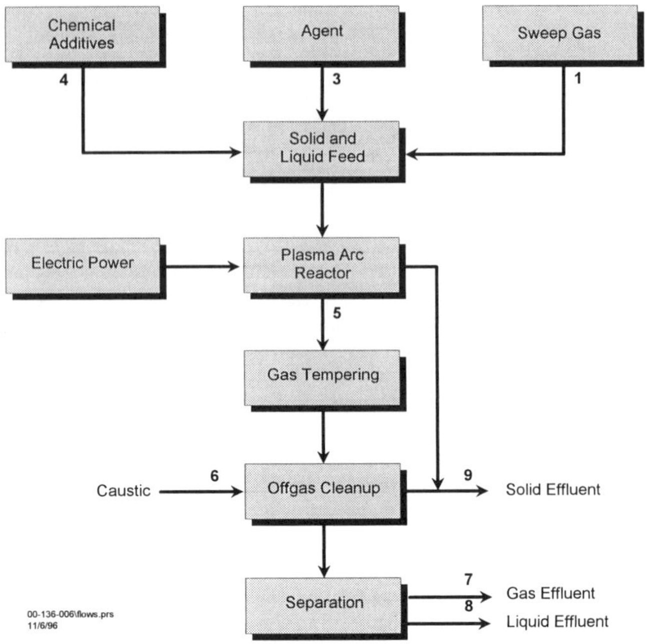

TABLE 3a. Carbon Electrode Technology[a] (per unit mass of chemical agent VX)

STREAM	1	3	4	5	6	7	8	9
Flow Components	Oxygen	Chemical Agent	Lime	Gas	Caustic Quench	Gas Effluent	Liquid Effluent	Solid Effluent
Fuel(Methane)								
Agent		1.0						
N_2	2.1		0.0	2.1		2.1		
O_2	0.6							
H_2				0.1		0.1		
H_2O					2.1	0.2	1.9	
CO				1.2		1.2		
NaOH					0.7		0.7	
CaO			0.4					
CaS								0.3
NaCN								
$Ca_2P_2O_7$								0.5
Other[a]						0.1		
TOTAL[b]	2.7	1.0	0.5	3.4	2.8	3.7	2.6	0.7

Total Input = 7.0 Total Output = 7.0

a This technology is not yet fully understood, so unknown products may have been involved that contributed hydrogen, nitrogen, and oxygen to the system.

b Round-off error causes some of the totals to be 0.1 lb higher/lower than the sum of all the constituents in some columns.

TABLE 3b. Carbon Electrode Technology (per unit mass of chemical agent GB)

STREAM	1	3	4	5	6	7	8	9
Flow Components	Oxygen	Chemical Agent	Lime	Gas	Caustic Quench	Gas Effluent	Liquid Effluent	Solid Effluent
Fuel(Methane)								
Agent		1.0						
N_2	1.5		0.0	1.5		1.5		
O_2	0.5							
H_2				0.1		0.1		
H_2O					1.4	0.1	1.3	
CO				0.8		0.8		
NaOH					0.5		0.3	
CaO			0.6					
CaF_2								0.3
NaF							0.2	
$Ca_2P_2O_7$								0.9
HF								
Other[a]						0.0		
TOTAL[b]	1.9	1.0	0.6	2.3	1.9	2.5	1.7	1.2

Total Input = 5.4 Total Output = 5.4

a This technology is not yet fully understood, so unknown products may have been involved that contributed hydrogen, nitrogen, and oxygen to the system.

b Round-off error causes some of the totals to be 0.1 lb higher /lower than the sum of all the constituents in some columns.

TABLE 3c. Carbon Electrode Technology (per unit mass of chemical agent HD)

STREAM	1	3	4	5	6	7	8	9
Flow Components	Oxygen	Chemical Agent	Lime	Gas	Caustic Quench	Gas Effluent	Liquid Effluent	Solid Effluent
Fuel(Methane)								
Agent		1.0						
N_2	0.8		0.1	0.8		0.8		
O_2	0.2							
H_2				0.1		0.1		
H_2O				0.1	1.2	0.1	1.2	
CO				0.7		0.7		
NaOH					0.4		0.4	
CaO			0.7					0.0
CaS								0.5
Na_2S							0.0	
Na_2CO_3							0.1	
$CaCl_2$								0.9
Other[a]					0.2			
TOTAL[b]	1.0	1.0	0.8	1.6	1.8	1.6	1.7	1.3

Total Input = 4.6 Total Output = 4.6

a This technology is not yet fully understood, so unknown reactants may have been involved that contributed carbon, chlorine, oxygen, hydrogen, sulfur, and calcium to the system.

b Round-off error causes some of the totals to be 0.1 lb higher /lower than the sum of all the constituents in some columns.

Figure 4. Plasma Torch Technology

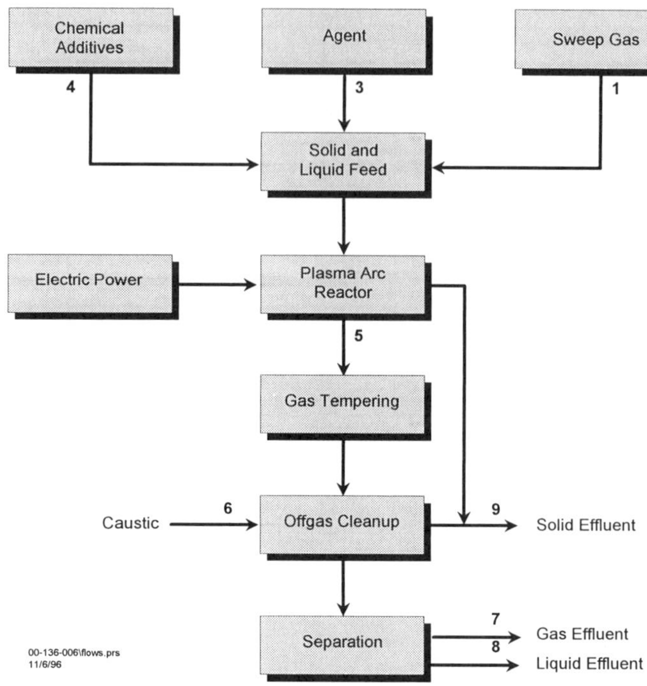

TABLE 4a. Plasma Torch Technology (per unit mass of chemical agent VX)

STREAM	1	3	4	5	6	7	8	9
Flow Components	Oxygen	Chemical Agent	Lime	Gas	Caustic Quench	Gas Effluent	Liquid Effluent	Solid Effluent
Fuel								
Agent		1.0						
N_2	2.1		0.0	2.1		2.1		
O_2	0.6							
H_2				0.1		0.1		
H_2O					2.1	0.2	1.9	
CO				1.2		1.2		
NaOH					0.7		0.7	
CaO			0.4					
CaS								0.3
NaCN								
$Ca_2P_2O_7$								0.5
Other[a]						0.1		
TOTAL[b]	2.7	1.0	0.5	3.4	2.8	3.7	2.6	0.7

Total Input = 7.0 Total Output = 7.0

a This technology is not yet fully understood, so unknown products may have been involved that contributed hydrogen, nitrogen, and oxygen to the system.

b Round-off error causes some of the totals to be 0.1 lb higher/lower than the sum of all the constituents in some columns.

TABLE 4b. Plasma Torch Technology (per unit mass of chemical agent GB)

STREAM	1	3	4	5	6	7	8	9
Flow Components	Oxygen	Chemical Agent	Lime	Gas	Caustic Quench	Gas Effluent	Liquid Effluent	Solid Effluent
Fuel								
Agent		1.0						
N_2	1.5		0.0	1.5		1.5		
O_2	0.5							
H_2				0.1		0.1		
H_2O					1.4	0.1	1.3	
CO				0.8		0.8		
NaOH					0.5		0.3	
CaO			0.6					
CaF_2								0.3
NaF							0.2	
$Ca_2P_2O_7$								0.9
HF								
Other[a]						0.0		
TOTAL[b]	1.9	1.0	0.6	2.3	1.9	2.5	1.7	1.2

Total Input = 5.4 Total Output = 5.4

a This technology is not yet fully understood, so unknown products may have been involved that contributed hydrogen, nitrogen, and oxygen to the system.

b Round-off error causes some of the totals to be 0.1 lb higher/lower than the sum of all the constituents in some columns.

TABLE 4c. Plasma Torch Technology (per unit mass of chemical agent HD)

STREAM	1	3	4	5	6	7	8	9
Flow Components	Oxygen	Chemical Agent	Lime	Gas	Caustic Quench	Gas Effluent	Liquid Effluent	Solid Effluent
Fuel								
Agent		1.0						
N_2	0.8		0.1	0.8		0.8		
O_2	0.2							
H_2				0.1		0.1		
H_2O				0.1	1.2	0.1	1.2	
CO				0.7		0.7		
NaOH					0.4		0.4	
CaO			0.7					0.0
CaS								0.5
Na_2S							0.0	
Na_2C_{O3}							0.1	
$CaCl_2$								0.9
Other[a]					0.2			
TOTAL[b]	1.0	1.0	0.8	1.6	1.8	1.6	1.7	1.3

Total Input = 4.6 Total Output = 4.6

a This technology is not fully understood, so unknown reactants may have been involved that contributed carbon, chlorine, oxygen, hydrogen, sulfur, and calcium to the system.

b Round-off error causes some of the totals to be 0.1 lb higher/lower than the sum of all the constituents in some columns.

Figure 5. Molten Salt Oxidation

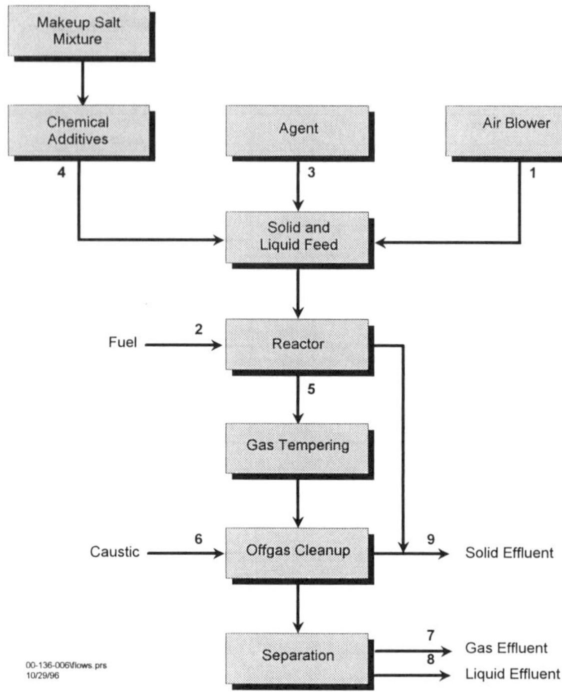

TABLE 5a. Molten Salt Oxidation (per unit mass of chemical agent VX)

STREAM	1	3	4	5	6	7	8	9
Flow Components	Air	Chemical Agent	Bed Makeup	Gas	Caustic Quench	Gas Effluent	Liquid Effluent	Solid Effluent
Fuel(Methane)								
Agent		1.0						
N_2	11.3			11.3		11.3		
O_2	3.2			1.2		1.2		
H_2O				0.9	12.9	10.0	3.8	
CO_2				2.2		2.2		
NaOH					0.4			
P_2O_5								
Na_2CO_3			1.0					0.2
Na_2HPO_4							0.4	
Na_2SO_3							0.4	
Other[a]							0.2	
Total[b]	14.5	1.0	1.0	15.5	13.3	24.7	4.9	0.2

Total Input = 29.8 Total Output =29.8

a This technology is not yet fully understood, so unknown products may have been involved that contributed sodium, nitrogen, hydrogen, and phosphorus to the system.

b Round-off error causes some of the totals to be 0.1 lb higher /lower than the sum of all the constituents in some columns.

TABLE 5b. Molten Salt Oxidation (per unit mass of chemical agent GB)

STREAM	1	3	4	5	6	7	8	9
Flow Components	Air	Chemical Agent	Bed Makeup	Gas	Caustic Quench	Gas Effluent	Liquid Effluent	Solid Effluent
Fuel(Methane)								
Agent		1.0						
N_2	7.5			7.5		7.5		
O_2	2.1			0.7		0.8		
H_2O				0.6	8.7	7.4	2.0	
CO_2				1.9		1.8		
NaOH					0.3			
P_2O_5								
Na_2CO_3			3.0					3.0
NaF							0.1	
Na_2HPO_4							0.2	
Other[a]			0.1					
Total[b]	9.6	1.0	3.1	10.8	9.0	17.5	2.2	3.0

Total Input = 22.7 Total Output = 22.7

a This technology is not yet fully understood, so unknown reactants may have been involved that contributed carbon and oxygen to the system.

b Round-off error causes some of the totals to be 0.1 lb higher /lower than the sum of all the constituents in some columns.

TABLE 5c. Molten Salt Oxidation (per unit mass of chemical agent HD)

STREAM	1	3	4	5	6	7	8	9
Flow Components	Air	Chemical Agent	Bed Makeup	Gas	Caustic Quench	Gas Effluent	Liquid Effluent	Solid Effluent
Na_2CO_3			1.4					1.4
Agent		1.0						
N_2	7.1			7.2		7.2		
O_2	2.0			0.8		0.8		
H_2O				0.5	7.8	7.1	1.2	
CO_2				1.7		1.6		
SO_2								
HCl								
NaOH					0.3			
NaCl							0.1	
Na_2SO_3							0.2	
Other[a]			0.1					
Total[b]	9.1	1.0	1.5	10.1	8.1	16.7	1.6	1.4

Total Input = 19.7 Total Output = 19.7

a This technology is not yet fully understood, so unknown reactants may have been involved that contributed carbon and oxygen to the system.

b Round-off error causes some of the totals to be 0.1 lb higher /lower than the sum of all the constituents in some columns.

Figure 6. Gas-Phase Reduction Process

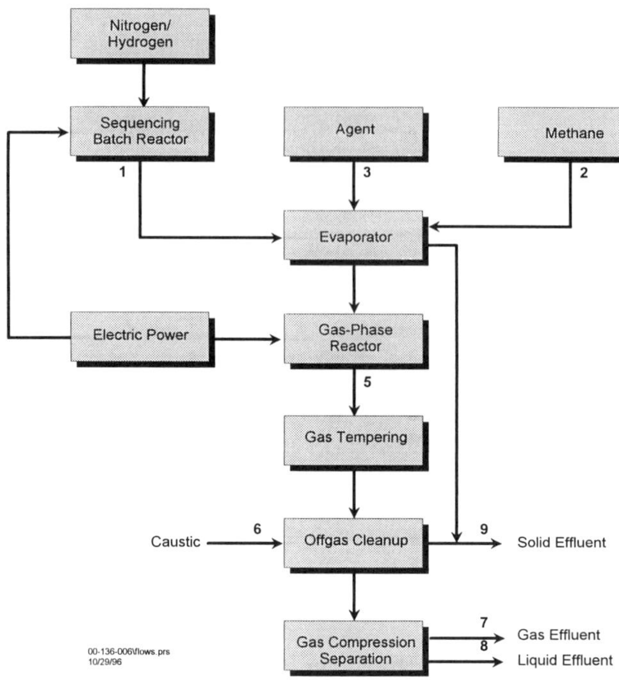

TABLE 6a. Gas-Phase Reduction Process (per unit mass of chemical agent VX)

STREAM	1	2	3	4	5	6	7	8	9
Flow Components	Nitrogen/ Hydrogen	Methane	Chemical Agent	Water	Gas Phase	Caustic	Gas Effl	Liquid Effl	Solid Effl
Methane		0.04							
Agent			1.0						
N_2	0.05				0.05		0.05		
CH_4					0.19		0.19		
H_2	0.19				0.04		0.04		
H_2O				0.53				0.46	
CO					0.02		0.02		
CO_2					0.12		0.12		
H_2S					0.11		0.11		
PH_3					0.14		0.14		
NH_3					0.06		0.06		
Other[a]									0.55
Total[b]	0.24	0.04	1.0	0.53	0.74		0.74	0.46	0.55

Total Input = 1.75 Total Output = 1.75

a This technology is not yet fully understood, so unknown products may have been involved that contributed hydrogen and nitrogen to the system.

b Round-off error causes some of the totals to be 0.1 lb higher/lower than the sum of all the constituents in some columns.

TABLE 6b. Gas-Phase Reduction Process (per unit mass of chemical agent GB)

STREAM	1	2	3	4	5	6	7	8	9
Flow Components	Nitrogen/ Hydrogen	Methane	Chemical Agent	Water	Gas Phase	Caustic	Gas Effl	Liquid Effl	Solid Effl
Methane		0.05							
Agent			1.0						
N_2	0.05				0.05		0.05		
CH_4					0.26		0.26		
H_2	0.14				0.06		0.06		
H_2O				0.50		0.29		0.43	
CO					0.03		0.02		
CO_2					0.17		0.17		
H_2S					0.21		0.21		
HCl					0.4				
NaOH						0.96			
Other[a]								1.79	
Total[b]	0.19	0.05	1.0	0.50	1.12	1.25	0.77	2.22	

Total Input = 2.99 Total Output = 2.99

a This technology is not yet fully understood, so unknown products may have been involved that contributed hydrogen and nitrogen to the system.

b Round-off error causes some of the totals to be 0.1 lb higher/lower than the sum of all the constituents in some columns.

Figure 7. Supercritical Water Oxidation

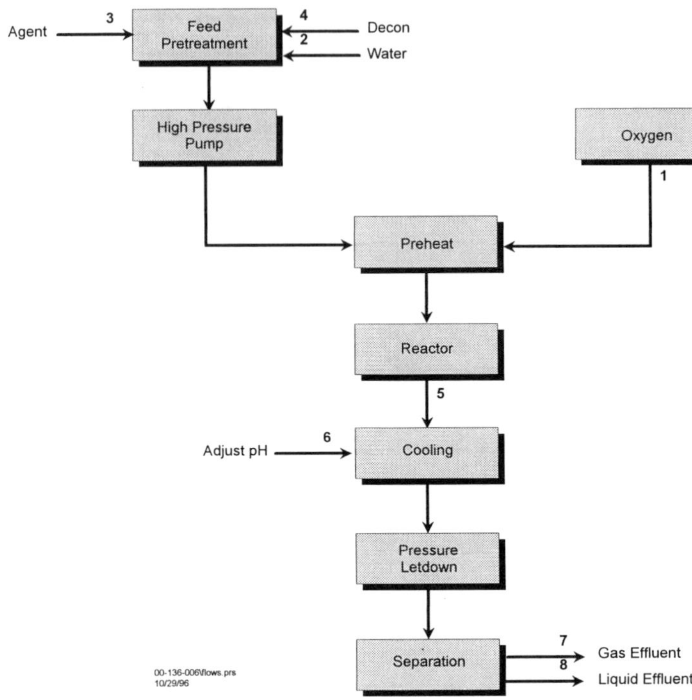

TABLE 7. Supercritical Water Oxidation (per unit mass of chemical agent VX)

STREAM	1	2	3	4	5	6	7	8
Flow Components	Oxygen	Water	Chemical Agent	Water & Decon	Reactor Effluent	Adjust pH	Gas Effluent	Liquid Effluent
Fuel(Methane)								
Agent			1.0					
N_2								
O_2	2.8				0.2		0.4	
H_2O		16.6		2.9	20.1	3.8		24.6
CO_2					2.6		1.9	
SO_2					0.2			
P_2O_5					0.3			
NaOH						0.6		
H_2SO_4								0.4
H_3PO_4								0.4
Other[a]				0.1				
Total[b]	2.8	16.6	1.0	3.0	23.3	4.4	2.3	25.5

Total Input = 27.8 Total Output = 27.8

a This technology is not yet fully understood, so unknown reactants may have been involved that contributed hydrogen and oxygen to the system.

b Round-off error causes some of the totals to be 0.1 lb higher/lower than the sum of all the constituents in some columns.

TABLE 7b. Supercritical Water Oxidation (per unit mass of chemical agent GB)

STREAM	1	2	3	4	5	6	7	8
Flow Components	Oxygen	Water	Chemical Agent	Water & Decon	Reactor Effluent	Adjust pH	Gas Effluent	Liquid Effluent
Fuel(Methane)								
Agent			1.0					
N_2								
O_2	1.7						0.3	
H_2O		16.0		1.9	17.8	2.6		20.8
CO_2					1.8		1.3	
SO_2								
P_2O_5					0.5			
NaOH						0.4		
HF					0.1			0.1
H_3PO_4								0.7
Other[a]				0.1				0.5
Total[b]	1.7	16.0	1.0	2.0	20.2	3.0	1.6	22.1

Total Input = 23.7 Total Output = 23.7

a This technology is not yet fully understood, so unknown reactants may have been involved that contributed carbon and oxygen to the system.

b Round-off error causes some of the totals to be 0.1 lb higher/lower than the sum of all the constituents in some columns.

TABLE 7c. Supercritical Water Oxidation (per unit mass of chemical agent HD)

STREAM	1	2	3	4	5	6	7	8
Flow Components	Oxygen	Water	Chemical Agent	Water & Decon	Reactor Effluent	Adjust pH	Gas Effluent	Liquid Effluent
Fuel(Methane)								
Agent			1.0					
N_2								
O_2	1.7				0.1		0.3	
H_2O		17.3		1.6	19.0	2.2		22.0
CO_2					1.5		1.1	
SO_2					0.5			
P_2O_5								
NaOH						0.3		
H_2SO_4								0.6
HCl					0.4			0.5
Other[a]				0.2				
Total[b]	1.7	17.3	1.0	1.9	21.5	2.6	1.4	23.1

Total Input = 24.5 Total Output = 24.5

a This technology is not yet fully understood, so unknown reactants may have been involved that contributed carbon and sulfur to the system

b Round-off error causes some of the totals to be 0.1 lb higher/lower than the sum of all the constituents in some columns

Figure 8. Wet Air Oxidation

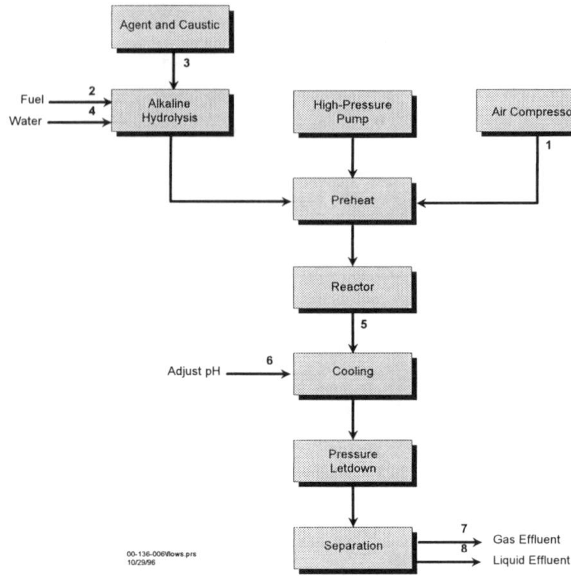

TABLE 8a. Wet Air Oxidation (per unit mass of chemical agent VX)

STREAM	1	2	3	4	5	6	7	8
Flow Components	Air	Fuel	Agent & Caustic	Decon & Water	Reactor Effluent	Adjust pH	Gas Effluent	Liquid Effluent
Fuel(Methane)		0.0						
Agent			1.0					
N_2	10.1				10.1		10.1	
O_2	2.6				0.4		0.4	
H_2O			4.0	17.7	23.5	0.0	1.5	22.3
CO_2					0.1		0.1	
$NaHCO_3$					2.8			2.8
NaOH			4.0		0.5			
H_2SO_4						2.2		
Na_2SO_4					3.6			3.7
$Na3PO_4$					0.6			0.6
NaOCl								
Other[a]				0.5	0.3		0.2	0.7
Total[b]	12.7	0.0	9.1	18.2	41.7	2.2	12.2	30.0

Total Input = 42.2 Total Output = 42.2

a This technology is not yet fully understood, so unknown reactants and products may have been involved that contributed carbon, nitrogen, oxygen, hydrogen, and sodium to the system.

b Round-off error causes some of the totals to be 0.1 lb higher/lower than the sum of all the constituents in some columns.

TABLE 8b. Wet Air Oxidation (per unit mass of chemical agent GB)

STREAM	1	2	3	4	5	6	7	8
Flow Components	Air	Fuel	Agent & Caustic	Decon & Water	Reactor Effluent	Adjust pH	Gas Effluent	Liquid Effluent
Fuel(Methane)		0.0						
Agent			1.0					
N_2	5.3				5.3		5.3	
O_2	1.6				0.2		0.2	
H_2O			3.5	10.9	15.9	0.0	0.9	15.0
CO_2					0.1		0.1	
Na_2CO_3					2.3			0.1
NaOH			3.5		0.4			
H_2SO_4						1.5		
Na_2SO_4								2.2
$Na3PO_4$					1.2			1.2
NaF					0.3			0.3
Other[a]				0.1	0.0		0.0	2.2
Total[b]	6.9	0.0	8.0	11.0	25.6	1.6	6.5	21.0

Total Input = 27.5 Total Output = 27.5

a This technology is not yet fully understood, so unknown reactants and products may have been involved that contributed carbon, oxygen, hydrogen, and sodium to the system.

b Round-off error causes some of the totals to be 0.1 lb higher/lower than the sum of all the constituents in some columns.

TABLE 8c. Wet Air Oxidation (per unit mass of chemical agent HD)

STREAM	1	2	3	4	5	6	7	8
Flow Components	Air	Fuel	Agent & Caustic	Decon & Water	Reactor Effluent	Adjust pH	Gas Effluent	Liquid Effluent
Fuel(Methane)		0.0						
Agent			1.0					
N_2	5.1				5.1		5.1	
O_2	1.5				0.4		0.2	0.1
H_2O			3.0	8.2	12.2	0.0	0.8	11.6
CO_2					0.1		0.1	
$NaHCO_3$								1.5
NaOH			3.0		0.4			
H_2SO_4						1.4		
Na_2SO_4					0.9			2.9
Na_2CO_3					2.0			0.1
NaCl					0.7			0.7
Other[a]	0.0			0.2	0.4			0.4
Total[b]	6.7	0.0	7.0	8.4	22.2	1.4	6.2	17.3

Total Input = 23.5 Total Output = 23.5

a This technology is not yet fully understood, so unknown reactants and products may have been involved that contributed carbon, oxygen, hydrogen, and sodium to the system.

b Round-off error causes some of the totals to be 0.1 lb higher/lower than the sum of all the constituents in some columns.

Figure 9. Steam Reforming

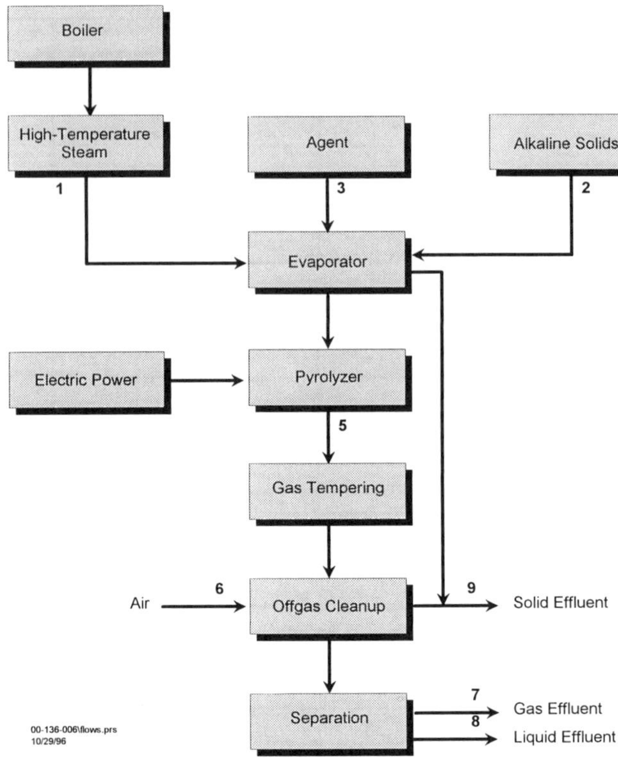

TABLE 9a. Steam Reforming (per unit mass of chemical agent VX)

STREAM	1	2	3	5	6	7	9
Flow Components	Steam	Alkaline Solids	Chemical Agent	Offgas	Air	Gas Effluent	Solid Effluent
NaOH		0.6					0.3
Agent			1.0				
N_2					8.8	8.8	
H_2				0.2			
H_2O	0.1			0.2		1.3	
CO				1.2			
CO_2						1.8	
P_2O_5							
Na_2SO_3							0.4
Na_2PO_3							0.4
O_2					2.4		
Other[a]		0.2					
Total[b]	0.1	0.8	1.0	1.5	11.2	11.9	1.2

Total Input = 13.1 Total Output = 13.1

a This technology is not yet fully understood, so unknown reactants may have been involved that contributed hydrogen and sodium to the system.

b Round-off error causes some of the totals to be 0.1 lb higher/lower than the sum of all the constituents in some columns.

TABLE 9b. Steam Reforming (per unit mass of chemical agent GB)

STREAM	1	2	3	5	6	7	9
Flow Components	Steam	Alkaline Solids	Chemical Agent	Offgas	Air	Gas Effluent	Solid Effluent
NaOH		0.9					
Agent			1.0				
N_2					6.2	6.2	
H_2				0.1			
H_2O	0.1			0.1		1.1	
CO				0.8			
CO_2						1.3	
P_2O_5							
NaF							0.3
Na_2PO_3							0.9
O_2					1.6		
Other[a]					0.0		
Total[b]	0.1	0.9	1.0	1.0	7.8	8.6	1.2

Total Input = 9.8 Total Output = 9.8

a This technology is not yet fully understood, so unknown reactants may have been involved that contributed hydrogen and carbon to the system.

b Round-off error causes some of the totals to be 0.1 lb higher/lower than the sum of all the constituents in some columns.

TABLE 9c. Steam Reforming (per unit mass of chemical agent HD)

STREAM	1	2	3	5	6	7	9
Flow Components	Steam	Alkaline Solids	Chemical Agent	Offgas	Air	Gas Effluent	Solid Effluent
NaOH		0.8					
Agent			1.0				
N_2					4.8	4.8	
H_2				0.1			
H_2O	0.1			0.6		0.6	
CO				0.7			
CO_2						1.0	
P_2O_5							
Na_2SO_3							0.8
NaCl							0.8
O_2					1.3		
Other[a]							0.1
Total[b]	0.1	0.8	1.0	1.3	6.1	6.4	1.6

Total Input = 8.0 Total Output = 8.0

a This technology is not yet fully understood, so unknown products may have been involved that contributed hydrogen and carbon to the system.

b Round-off error causes some of the totals to be 0.1 lb higher/lower than the sum of all the constituents in some columns.

Figure 10. Catalytic Oxidation as Secondary Combustion Process

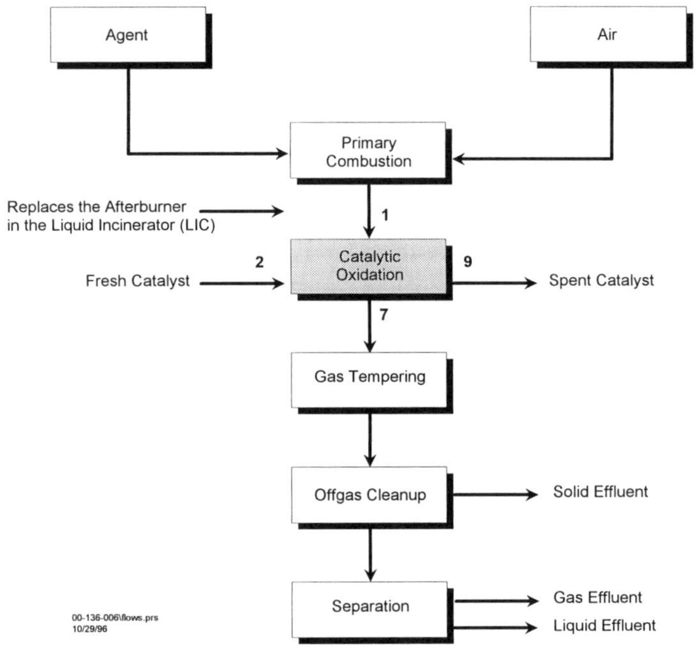

TABLE 10a. Catalytic Oxidation (per unit mass of chemical agent VX)

STREAM	1	2	7	9
Flow Components	Primary Flue Gas	Fresh Catalyst	Gas Effluent	Solid Effluent
Catalyst		0.1		0.1
Agent				
N_2	13.1		13.1	
O_2	1.7		1.7	
H_2O	3.1		3.1	
CO_2	1.8		1.8	
SO_2	0.2		0.2	
P_2O_5	0.2		0.2	
Na_2SO_3				
Na_2HPO_4				
NaOH				
TOTAL[a]	20.0	0.1	20.0	0.1
	Total Input = 20.1		Total Output = 20.1	

a Round-off error causes some of the totals to be 0.1 lb higher/lower than the sum of all the constituents in some columns.

TABLE 10b. Catalytic Oxidation (per unit mass of chemical agent GB)

STREAM	1	2	7	9
Flow Components	Primary Flue Gas	Fresh Catalyst	Gas Effluent	Solid Effluent
Catalyst		0.1		0.1
Agent				
N_2	12.6		12.6	
O_2	1.5		1.5	
H_2O	3.1		3.1	
CO_2	1.8		1.8	
HF	0.1		0.1	
P_2O_5	0.5		0.5	
NaOH				
NaF				
Na_2HPO_4				
TOTALa	19.7	0.1	19.7	0.1

Total Input = 19.8 Total Output = 19.8

a Round-off error causes some of the totals to be 0.1 lb higher/lower than the sum of all the constituents in some columns.

TABLE 10c. Catalytic Oxidation (per unit mass of chemical agent HD)

STREAM	1	2	7	9
Flow Components	Primary Flue Gas	Fresh Catalyst	Gas Effluent	Solid Effluent
Catalyst		0.1		0.1
Agent				
N_2	12.6		12.6	
O_2	1.5		1.5	
H_2O	2.9		2.9	
CO_2	1.8		1.8	
SO_2	0.6		0.6	
HCl	0.5		0.5	
NaOH				
NaCl				
Na_2SO_3				
TOTALa	19.8	0.1	19.8	0.1

Total Input = 19.9 Total Output = 19.9

a Round-off error causes some of the totals to be 0.1 lb higher/lower than the sum of all the constituents in some columns.

Figure 11. Reaction with Sulfur

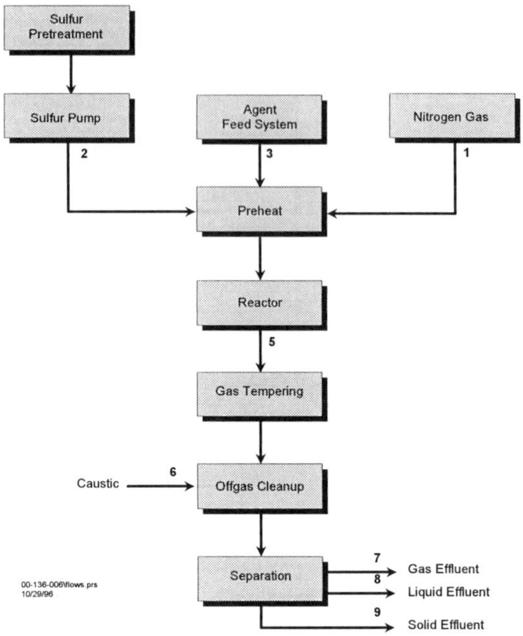

TABLE 11a. Reaction with Sulfur (per unit mass of chemical agent VX)

STREAM	1	2	3	5	6	7	8	9
Flow Components	Nitrogen	Sulfur	Chemical Agent	Flue Gas	Water & Oxygen	Gas Effluent	Liquid Effluent	Solid Effluent
Fuel(Methane)								
Agent			1.0					
N_2	0.0			0.0		0.0		
S		0.7						
H_2O					0.3		0.3	
H_2S				1.6				
CS								0.9
SO_2				0.3				
O_2					0.7			
Na_2SO_3							0.4	
Na_2S								
P_2S5								0.4
Other[a]						0.7		
Total[b]	0.0	0.7	1.0	1.9	1.0	0.7	0.7	1.3

Total Input = 2.7 Total Output = 2.7

a This technology is not yet fully understood, so unknown reactants may have been involved that contributed carbon, hydrogen, and oxygen to the system.

b Round-off error causes some of the totals to be 0.1 lb higher/lower than the sum of all the constituents in some columns.

TABLE 11b. Reaction with Sulfur (per unit mass of chemical agent GB)

STREAM	1	2	3	5	6	7	8	9
Flow Components	Nitrogen	Sulfur	Chemical Agent	Flue Gas	Water & Oxygen	Gas Effluent	Liquid Effluent	Solid Effluent
HF				0.1				
Agent			1.0					
N_2	0.0			0.0		0.0		
O_2					0.5			
S		1.1						
H_2O					0.9		0.9	
H_2S				1.1				
CS								0.6
SO_2				0.2				
NaOH								
Na_2SO_3							0.9	
NaF							0.3	
P_2S5								0.2
Other[a]						0.7		
Total[b]	0.0	1.1	1.0	1.4	1.4	0.7	2.0	0.8

Total Input = 3.5 Total Output = 3.5

a This technology is not yet fully understood, so unknown products may have been involved that contributed carbon, hydrogen, and oxygen to the system.

b Round-off error causes some of the totals to be 0.1 lb higher/lower than the sum of all the constituents in some columns.

TABLE 11c. Reaction with Sulfur (per unit mass of chemical agent HD)

STREAM	1	2	3	5	6	7	8	9
Flow Components	Nitrogen	Sulfur	Chemical Agent	Flue Gas	Caustic & Oxygen	Gas Effluent	Liquid Effluent	Solid Effluent
O_2					0.3			
Agent			1.0					
N_2	0.0			0.0		0.0		
S		0.0						
H_2O					1.7		2.3	
H_2S				0.6				
CS								0.5
HCl				0.4				
NaOH					0.6			
NaCl							0.8	
Na_2S								
Na_2SO_3							0.2	
Other[a]					0.1			
Total[b]	0.0	0.0	1.0	1.0	2.8	0.0	3.3	0.5

Total Input = 3.8 Total Output = 3.8

a This technology is not yet fully understood, so unknown products may have been involved that contributed carbon, hydrogen, and oxygen to the system.

b Round-off error causes some of the totals to be 0.1 lb higher/lower than the sum of all the constituents in some columns.

Figure 12. Gasification Process

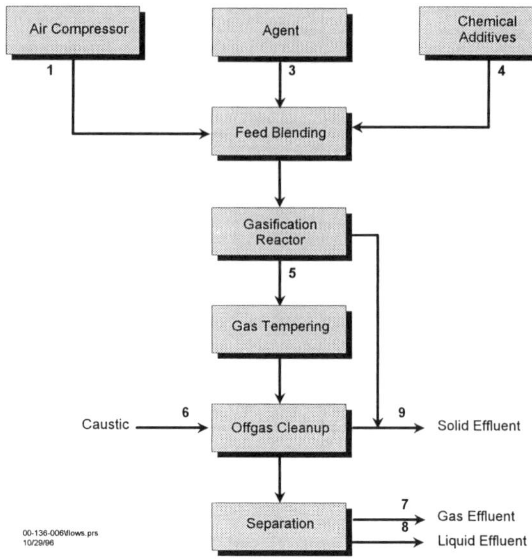

TABLE 12a. Gasification Process (per unit mass of chemical agent VX)

STREAM	1	3	4	5	6	7	8	9
Flow Components	Optional Air	Chemical Agent	Optional Lime	Flue Gas	Quench& Caustic	Gas Effluent	Liquid Effluent	Solid Effluent
Fuel(Methane)								
Agent		1.0						
N_2	2.1		0.0	2.1		2.1		
O_2	0.6							
H_2				0.1		0.1		
H_2O					2.1	0.2	1.9	
CO				1.2		1.2		
NaOH					0.7		0.7	
CaO			0.4					
CaS								0.3
NaCN								
$Ca_2P_2O_7$								0.5
Other[a]							0.1	
Total[b]	2.7	1.0	0.5	3.4	2.8	3.5	2.8	0.7

Total Input = 7.0 Total Output = 7.0

a This technology is not yet fully understood, so unknown products may have been involved that contributed nitrogen, hydrogen, and oxygen to the system.

b Round-off error causes some of the totals to be 0.1 lb higher/lower than the sum of all the constituents in some columns.

TABLE 12b. Gasification Process (per unit mass of chemical agent GB)

STREAM	1	3	4	5	6	7	8	9
Flow Components	Optional Air	Chemical Agent	Optional Lime	Flue Gas	Quench & Caustic	Gas Effluent	Liquid Effluent	Solid Effluent
Fuel(Methane)								
Agent		1.0						
N_2	1.5		0.0	1.5		1.5		
O_2	0.5							
H_2				0.1		0.1		
H_2O					1.4	0.1	1.3	
CO				0.8		0.8		
NaOH					0.5		0.5	
CaO			0.6					
CaF2								0.3
NaF							0.2	
$Ca_2P_2O_7$								0.9
HF								
Other[a]					0.3			
Total[b]	1.9	1.0	0.6	2.3	2.1	2.5	1.9	1.2

Total Input = 5.6 Total Output = 5.6

a This technology is not yet fully understood, so unknown reactants may have been involved that contributed sodium and fluorine to the system.

b Round-off error causes some of the totals to be 0.1 lb higher/lower than the sum of all the constituents in some columns.

TABLE 12c. Gasification Process (per unit mass of chemical agent HD)

STREAM	1	3	4	5	6	7	8	9
Flow Components	Optional Air	Chemical Agent	Optional Lime	Flue Gas	Quench& Caustic	Gas Effluent	Liquid Effluent	Solid Effluent
Fuel(Methane)								
Agent		1.0						
N_2	0.8		0.1	0.8		0.8		
O_2	0.2							
H_2				0.1		0.1		
H_2O					1.2	0.1	1.2	
CO				0.7		0.7		
NaOH					0.4		0.3	
CaO			0.7					0.1
CaS								0.5
Na_2S							0.0	
Na_2CO_3							0.1	
$CaCl_2$								0.9
Other[a]					0.1			
Total[b]	1.0	1.0	0.8	1.5	1.8	1.6	1.6	1.4

Total Input = 4.6 Total Output = 4.6

a This technology is not yet fully understood, so unknown reactants may have been involved that contributed carbon, hydrogen, sulfur, calcium, chlorine, and oxygen to the system.

b Round-off error causes some of the totals to be 0.1 lb higher/lower than the sum of all the constituents in some columns.

Figure 13. Caustic Hydrolysis

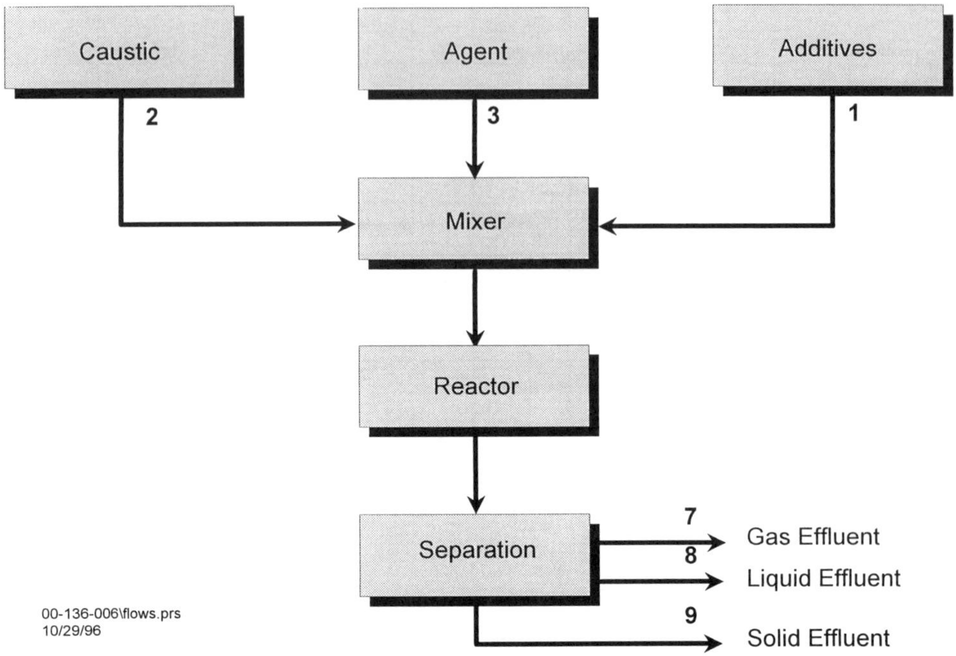

TABLE 13a. Caustic Hydrolysis (per unit mass of chemical agent VX)

STREAM	1	2	3	7	8
Flow Components	Air	Caustic & Hypochlorite	Agent	Gaseous Effluent	Liquid Effluent
Agent			1.0		
Air	1.73			1.73	
NaOCl		0.74			
NaOH		0.41			0.02
H_2O		8.15			7.63
NaCl		0.74			0.67
Sulfonate					1.05
EMPA & MPA					0.68
Other[a]					1.00
Total[b]	1.73	10.04	1.0	1.73	11.04

Total Input = 6.0 Total Output = 6.0

a This technology is not yet fully understood, so unknown reactants may have been involved that contributed sodium to the system. ... Ref. U.S. Army NECA Concept Design, May 1996

b Round-off error causes some of the totals to be 0.1 lb higher/lower than the sum of all the constituents in some columns.

TABLE 13b. Caustic Hydrolysis (per unit mass of chemical agent GB)

STREAM	2	3	8
Flow Components	Caustic	Agent	Liquid Effluent
Agent		1.0	
NaOH	0.6		0.0
H_2O	2.3		2.3
Phosphonate			1.1
NaF			0.3
Other[a]			0.2
Total[b]	2.9	1.0	3.9

Total Input = 3.9 Total Output = 3.9

a This technology is not yet fully understood, so unknown products may have been involved that contributed carbon, hydrogen, and oxygen to the system.

b Round-off error causes some of the totals to be 0.1 lb higher/lower than the sum of all the constituents in some columns.

Figure 14. Water Hydrolysis

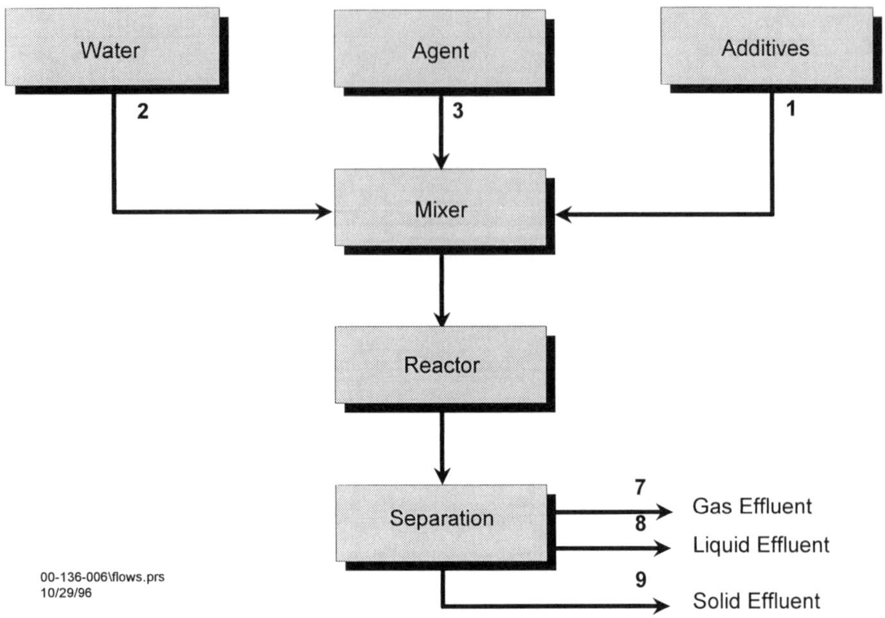

TABLE 14. Water Hydrolysis With Offsite Treatment (per unit mass of chemical agent HD)

STREAM	2	3	7	8
Flow Components	Caustic & Water	Agent	Gaseous Effluent	Liquid Effluent
Agent		1.0		
Caustic	0.50			
O_2	0.38		0.38	
N_2	1.19		1.27	
CO_2	0.08			
H_2O	28.03			28.03
Thiodiglycol				0.62
NaCl				0.71
Other				0.17
TOTALa	30.18	1.0	1.65	29.53

Total Input = 31.18 Total Output = 31.18

a Round-off error causes some of the totals to be 0.1 lb higher/lower than the sum of all the constituents in some columns.
Ref. U.S. Army APG Concept Design, May 1996

Figure 15. Biodegradation Following Water Hydrolysis

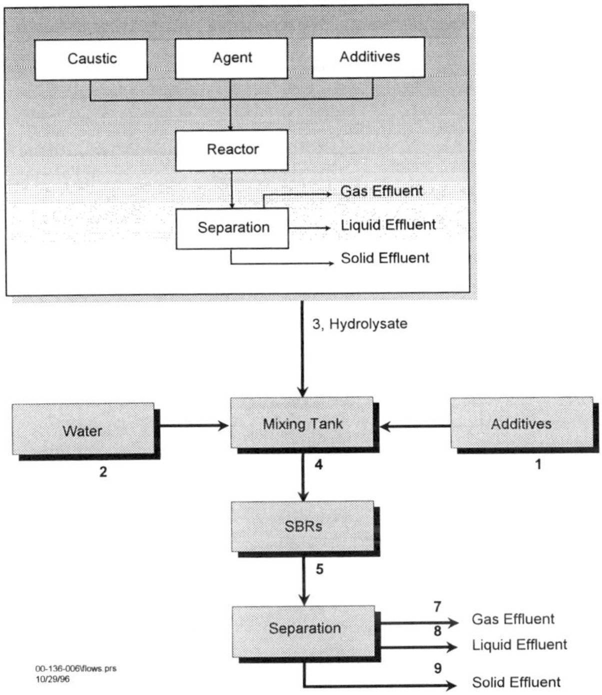

TABLE 15. Biodegradation of HD Following Water Hydrolysis (per unit mass of chemical agent HD)

STREAM	1	2	3	7	8
Flow Components	Air	Water	Hydrolysate	Gaseous Effluent	Liquid Effluent
Hydrolysate			29.53*		
Caustic					
O_2	15.73			15.73	
N_2	48.97			48.97	
CO_2	3.24			3.24	
H_2O		61.36			86.29
$NaHCO_3$		1.01			
NaCl					0.75
Other					5.86
TOTALa	67.94	62.37	29.53	67.94	92.90

Total Input = 160.84 Total Output = 160.84

a Round-off error causes some of the totals to be 0.1 lb higher/lower than the sum of all the constituents in some columns.

* Produced by hydrolyzing 1 pound of HD, Ref. U.S. Army APG Concept Design, May 1996

Figure 16. Electrochemical Oxidation With Silver

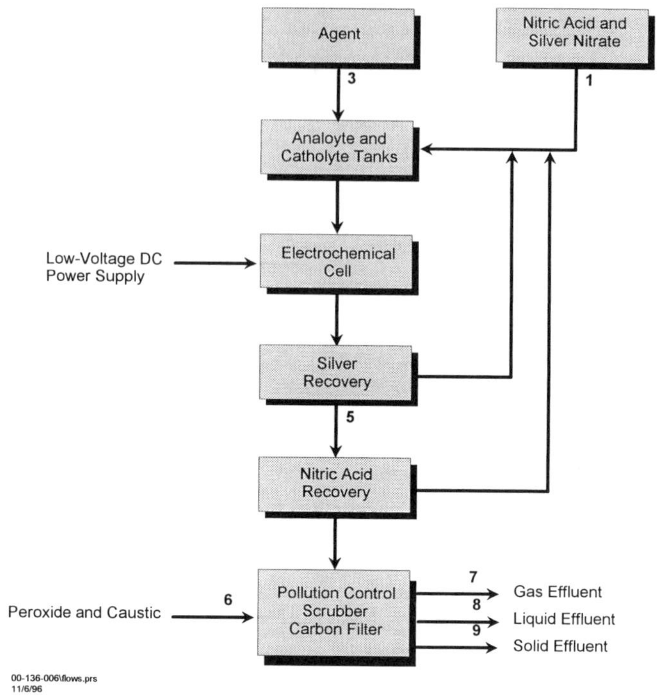

TABLE 16a. Silver 2, Electrochemical Oxidation (per unit mass of chemical agent VX)

STREAM	1	2	3	4	5	6	7	8	9
Flow Components	Nitric Acid	Nitrogen/ Oxygen	Chemical Agent		Offgas	Peroxide/ Caustic	Gas Effluent	Acid Waste	Brine Effluent
HNO_3	0.30							0.05	
Agent, VX			1.00						
$AgNO_3$									
N_2		0.25			0.25		0.25		
O_2		2.60			4.65		4.65		
H_2O						21.45		23.65	3.75
H_2O_2						9.60			
CO_2					1.80		1.80		
NOx					0.02		0.02		
NaOH						3.85			
$NaNO_3$									1.65
Na_2SO_4									1.30
NaCl									1.30
Other									0.63
TOTALa	0.30	2.85	1.00		6.72	34.90	6.72	23.70	8.63

Total Input = 39.05 Total Output = 39.05

a Round-off error causes some of the totals to be 0.1 lb higher/lower than the sum of all the constituents in some columns.

TABLE 16b. Silver 2, Electrochemical Oxidation (per unit mass of chemical agent HD)

STREAM	1	2	3	4	5	6	7	8	9
Flow Components	Nitric Acid	Nitrogen/ Oxygen	Chemical Agent		Offgas	Peroxide/ Caustic	Gas Effluent	Acid Waste	Brine Effluent
HNO_3	0.55							0.54	
Agent, VX			1.00						
$AgNO_3$									
N_2		0.15			0.15		0.15		
O_2		1.70			2.85		2.85		
H_2O						11.50		13.30	2.85
H_2O_2						5.40			
CO_2					1.10		1.10		
NOx					0.01		0.01		
NaOH						2.80			
$NaNO_3$									1.50
Na_2SO_4									1.70
NaCl									0.10
Other									
TOTALa	0.55	2.85	1.00		4.11	19.70	4.11	13.84	6.15

Total Input = 24.10 Total Output = 24.10

a Round-off error causes some of the totals to be 0.1 lb higher/lower than the sum of all the constituents in some columns.

REFERENCES

Aghamohammadi [1993] B., Durai-Swamy, K., Das, A.K., and Mansour, M.N., *Chemical Recovery by Gasification of Non-Wood Pulp Liquors*, Tappi Pulping Conference, Atlanta, GA. November 1993.

Aghamohammadi [1995] B., *Test Report for Sandia National Laboratory, MTCI Project No. 10075*. March 1995.

Ashland [1995] Petroleum Company, *HYMELT™, Molten Baths for Recycling Elements*, Company Literature.

Ball [1992] C.E., Randall, D., Huffman, G. and Lee, C.C., *Design and Testing of a Solar/Thermal Reactor to Destroy Hazardous Organic Compounds*, Proceedings of the 1992 Incineration Conference, Albuquerque, NM. May 11-15, 1992.

Binder [1993] W.D., Martin, R.J., Smythe, R.E., *Elimination of Refinery VOCs and HAPs Using Flameless thermal Oxidation*, Proceedings of the 1993 Incineration Conference, Knoxville, TN. May 4, 1993.

Bostick [1994] W.D. et al., *Surrogate Formulations for Thermal Treatment of Low-Level Mixed Waste, Part II: Selected Mixed Waste Treatment Project Waste Streams*, DOE/MWIP-16, Oak Ridge K-25 site, Oak Ridge, TN. January 1994.

Chan, S.H., *Multiphase Turbulent Liquid Metal Fuel Combustion,* Progress in Energy Combustion Science. Vol. 19, pp. 105-143, 1993.

Chen, T.H., C. Cambell, Y.L. Liang, K. Ward, and J. White, Reaction of RDX, HMX, TNT, and PETN with Sulfur, Report No. SMCAR-AEE-WE-93-26, Analytical Research and Services Explosives Section, Explosives and Warheads Branch, Picatinny Arsenal, NJ. October 1993.

Chemical Engineering Progress [1995], *Supercritical Water Oxidation Aims for Wastewater Cleanup*. pp. 10-18, April 1995.

Clark [1994], R.K., Pusateri, J.F, Lherbier, L.W., *Gas-Fired Flash Smelting for Metals Recovery from Hazardous Solid Wastes*, Proceedings of the Industrial Energy Technology Conference, Houston, TX. April 1994.

Crosley [1995] S. et al., *Review Comments on Draft Alternative to Incineration Area Status Report*, Oak Ridge National Laboratory, Oak Ridge, TN. February 8, 1995.

Cudahy [1995] J., Escarda, T., Gimpel, R., Huffman, G., Kolts, J., Musgrave, D., Peters, R. Seeker, R., Swartz, V., Torbert, M., Vavruska, J., *Integrated Thermal Treatment Systems Study, U. S. Department of Energy Internal Review Panel Report*, Office of Technology Development, Washington, DC. April 1995.

Cudahy [1993] J., Escarda, T., Graham, F.R., Helt, J.E., Martin, D.M., Thurnau, R.C., Seeker, W.R., Schwartz, V., Technical Review of Molten Salt Oxidation, U.S. Department of Energy, Office of Environmental Management, DOE/EM-0139P, Washington, DC. December 1993.

Davenport [1995] R.L., James, B, *Solar Soil Detoxification System Design*, Science Applications International Corp., Golden, CO. August 199

Dighe [1991] S.V., Paluson, D.K. and Lahoda, E.J., *Westinghouse Plasma Process for Treatment of Industrial Waste*, 1991 AISE Annual Convention, Pittsburgh, PA. September 23-26, 1991.

Department of Defens [1995], Advances in Alternative Demilitarization Technologies, Workshop at Reston, VA. 25-27 September 1995.

Energy and Environmental Research Corporation [1992], *Program Review for EPA Cooperative Agreement No. CR816832*, Cincinnati, OH. March 10, 1992.

Engleman [1995], V. and Holm, F.W., *Oxidation with Ultraviolet Light and Oxidizing Agents*, U.S. Army, SAIC DO-136, Aberdeen Proving Ground. May 1995.

Eschenbach [1993] R.C., Schleinger, M.E., Schleinger, M.P., *Waste Minimization with Plasma Processing*, Waste Management '93 Conference, Tucson, AZ. February 28-March 4, 1993.

Feizollahi [1995] F., Quapp, W.J., *Integrated Thermal Treatment System Study - Phase 2 Results*, INEL-95/0129, Idaho National Engineering Laboratory, Idaho Falls, ID. March 1995.

Galloway [1994] T.R. and Depetris, S.S., *Destroying LLW and Hazardou Waster On-Site with the Synthetica Steam Reformer*, Waste Management '94 Conference, Tucson, AZ. March-February 1994.

Galloway [1993] T.R. and Sprung, J.L., *Destruction of UST Organics and Nitrates, Polymeric Organic Wastes, and Cholorcarbon Solvents by Steam Reforming*, Proceedings of the Information Exchange Meeting on Waste Retrieval, Treatment and Processing, Houston, TX. March 15-17, 1993.

Gay [1991] R.L., Buried Waste Demonstration Technology Status Report for Molten Salt Waste Oxidation Process, TTP No. SF150601, Energy Technology Center, Rockwell International, Canoga Park, CA. December 20, 1991.

Glatzmaier [1990] G.C., Mehos, M.S., Nix, R.G., 1990, *Solar Destruction of Hazardous Chemicals*, Paper presented at 1990 ASME International Solar Energy Conference, Golden, CO. 1990.

Glatzmaier [1993] G.C. and Bohn, M.S., *Solar Assisted Combustion of 1,2-Dichlorobenzene*, AIChE/ASM National Heat Transfer Conference, Atlanta, GA. August 1993.

Gloyna [1993], E.E. and Li, L., 1993, *Supercritical Water Oxidation: An Engineering Update*, Waste Management. Vol 13, pp. 379-394, 1993.

Gorman [1996], P., and Bohn, M.S., *On-Sun Testing of A Solar Detoxification Reactor*, Proposed Paper for 1996 ASME International Solar Energy Conference, San Antonio, TX. March 31 - April 3, 1996.

Holm, F.W., *Survey and Update of Alternative Demilatarization Technologies*, Final Report, U.S. Army, SAIC DO-136, Aberdeen Proving Ground. January 1995.

Holm, F.W., *Trip Report on Base Catalyzed Decomposition*, Wright State University, SAIC DO-136. October 14, 1994.

Holm, F.W. et al., *Innovative Site Remediation Technology - Thermal Destruction*, Pending Publication by the American Academy of Environmental Engineers, Annapolis, MD. September 1996.

Holm, F.W. et al., *Innovative Site Remediation Technology - Thermal Destruction*, Published by the American Academy of Environmental Engineers, Annapolis, MD. June 1994.

Holm, F.W. et al., *Advances in Alternative Demilatarization Technologies*, pp. 317-346, U.S. Army Workshop Proceedings, SAIC DO-136, Reston, VA. January 1996.

Holm, F.W. et al., *Scientific Advances in Alternative Demilatarization Technologies*, pp. 111-128, NATO ASI Series, Kluwer Academic Publishers. June 1996.

Jarosch [1994], T.R., Haselow, J.S., Rossabi, J., Burdick, S,A., Raymond, R., Young, J.E. and Lombard, K.H., *Final Report on Testing of Off-Gas Treatment Technologies for Abatement of Atmospheric Emissions of Chorinated Volatile Organic Compounds (U)*, Westinghouse Savannah River Company, WSRC RP-94-927. September, 1994.

Joyce [1993] E.L., Lally, B., Ozturk B., Fruehan, R.J., *Liquid Metal Recycle and Waste Treatment: Liquid Metal Melt-Slag Technology Evaluation for MWIP*, TTP-AL132001, LAUR-93-3026, Los Alamos National Laboratory, Los Alamos, NM.

Lightfoot [1992], B.W., Robilliard, K.R., Matusewicz, R.W., *Ausmelt Submerged Lancing Technology for Smelting and Detoxification of Hazardous and Radioactive Waste*, Proceedings of the 1992 Incineration Conference, Albuquerque, NM. pp. 771-778, May 11-15, 1992.

Longwell, J. et al., *Alternative Technologies for the Destruction of Chemical Agents and Munitions,* Committee on Alternative Chemical Demilitarization Technologies, National Research Council. National Academy Press: Washington, D.C. June 1993.

Mensinger [1991] M.C., Rehmat, A., Bryans, B.G., Lau, F.S., Shearer, T.L. and Duggan, P.A., 1991, Experimental Development of a Two-Stage Fluidized Bed/Cyclonic Agglomerating Incinerator, Incineration Conference Proceeding, pp 603-612, Knoxville, TN. May 13-17, 1991.

Persoon [1994] J.R., Letter from Alliant TechSystems. September 28, 1994.

Schwinkendorf [1995] W.E., McFee, J., Devarakonda, M., Nenninger, L.L., Fadullon, F.S., Donaldson, T.L., Dickerson, K., *Alternatives to Incineration Technical Area Status Report*, DOE/MWIP-26, U.S. Department of Energy, Washington, DC 20585 April 1995.

Sheridan [1993] J.H., *Technology of the Year: Molten Metal Technology, Inc.*, Industry Week. December 20, 1993.

Malinka [1991] J.G., *Liquid Stabilized Plasma for Waste Treatment*, Proceedings of the 1991 Incineration Conference, Knoxville, TN. May 13-17, 1991.

Rehmat, A.G., Mensinger, M.C., *Two-Stage Fluidized-Bed/Cyclonic Agglomerating Combustor*, Final Report to EPA for the Period July 1, 1990 through June 30, 1994, EPA Contract No. CR-816911-01-3. July 1995.

Soelberg [1994] N.R. et al., *Arc Melter Demonstration Baseline Test Results*, EGG-WTD-11138, Idaho Nuclear Engineering Laboratory. July 1994.

Stelman [1992] D., Stewart, A.E., Yosim, S.J., and Gray, R.L., Treatment of Mixed Waste by the Molten Salt Oxidation Process, Proceedings of the 1992 Incineration Conference, Albuquerque, NM, pp. 795-799. May 11-15, 1992.

Surma [1993] J.E. et al. *Graphite Electrode dc Arc Technology Development for Treatment of Buried Wastes*, Proceedings of the 1993 Incineration Conference,

Knoxville, TN, pp. 453-456. May 3-7, 1993.
Task Force [1995] *Molten Salt Task Force Meeting*, Alberta, Canada. January 13-15, 1993.

Taylor [1992] G., Telephone conversation with Energy and Environmental Research Corporation, Irvine, CA. April 17, 1992.

Tester [1991] J.W. , Holgate, F.J., Armellinin, P.A., Webley, W.R., Killilea, G.T. Hong, and Barner, H.E., 1991, Supercritical Water Oxidation Technolgy: *A Review of Process Development and Fundamental Research*, ABCS Symposium Series on Emerging Technologies for HazardousWaste Management, Atlanta, GA. October 1-3, 1991.

Thermatix [1992] Inc., *Destruction of Organic compounds in the Thermatrix Flameless Thermal Oxidizer*, San Jose, CA. 1992.

Thompson, P.L., Peterson, D.R. and Holm, F.W., *Sreening and Evaluation Criteria for Alternative Demilitarization Technologies*, Final Report, U.S. Army, SAIC DO-136, Aberdeen Proving Grounds. August 1995.

Trescott [1995] J., Wilver, P., Hamilton, R.A., Wittle, J.K., *DC Graphite Arc Furnace, A Simple System to Reduce Mixed Waste Volume*, 17th Annual U.S. Department of Energy low-Level Radioactive Waste Management Conference, Phoenix, AZ. December 12-14, 1995.

University of Pittsburgh Applied Research Center, National Environmental Technology Applications Corporation and the Center for Hazardous Materials Research, Continuous Pilot-Scale Production Unit Demonstration of the Adams Process, Pittsburgh, PA. August 1991.

U.S. Army [May, 1996], *Aberdeen Proving Ground Concept Design Package for HD Neutralization followed by Biodegradation*, Program Manager for Chemical Demilitarization, Aberdeen Proving Ground. May, 1996.

U.S. Army [May, 1996], *Newport Chemical Activity Concept Design Package for VX Neutralization followed by Offsite Biodegradation,* Program Manager for Chemical Demilitarization, Aberdeen Proving Ground. May 1996.

U.S. Department of Energy [1995], Office of Environmental Management Technology Development, *Mixed Waste Characterization, Treatment, and Disposal Focus Area Technology Summary*, DOE/EM-0252. June 1995.

U.S. Department of Energy [1994a], *Supercritical Water Oxidation Program (SCWOP)*, DOE/EM-0121P, U.S. Department of Energy, Washington, DC 20585. February 1994.

U.S. Department of Energy [1994b], *Buried Waste Integrated Demonstration*, DOE/EM-0149P, Office of Technology Development, Washington, DC. March 1994.

U.S. Department of Energy [1994c], Office of Technology Development Preliminary Technology Fact Sheet, *Steam Plasma Torch Reforming Waste Destruction System Coupled with Gas Flare*, Washington, DC. March 7, 1994.

U.S. Department of Energy [1993], *EM-30 Mixed Waste Technology Review*, Lawrence Livermore National Laboratory (LLNL). July 8, 1993.

U.S. Environmental Protection Agency [1994], *Minipilot Solar System: Design/ Operation of System and Results of Non-Solar Testing at MRI*, Project Summary, EPA/600/SR-94/000, Risk Reduction Engineering Laboratory, Cincinnati, OH 45268. January 1994.

U.S. Environmental Protection Agency [1993], Superfund Innovative Technology Evaluation Emerging Technology Bulletin, *Spouted Bed Reactor*, EPA/540/F-93/508, August 1993.

U.S. Environmental Protection Agency [1991], The Superfund Innovative Technology Evaluation Program Technology Profiles, 4th Edition, EPA/540/5-91/008, Washington, D.C. November 1991.

Wilbourn [1994] R.G., Newburn, J.A. and Schofield, J.T., *Treatment of Hazardous Wastes Using the Thermatix Treatment System*, Proceedings of the 1994 Incineration Conference, Houston, TX, May 9-13, 1994.

Woods [1993] K.B. and Scholfield, J.T., *Control of Toxic air Emissions with a Flameless Thermal Oxidizer*, Air and Waste Management Association 86th Annual Meeting, Denver, CO. June 13-18, 1993.

Wright, L.E., Letter of October 17, 1995, Adams Process Testing at Pine Bluff Arsenal, Department of the Army, Pine Bluff Chemical Activity, Pine Bluff, AR.

Index